Engineering Health
How Biotechnology Changed Medicine

W0112675

Print ISBN: 978-1-78262-084-6
EPUB eISBN: 978-1-78801-234-8

A catalogue record for this book is available from the British Library

© The Royal Society of Chemistry 2018

The Royal Society of Chemistry is a charity, registered in England and Wales, Number 207890, and a company incorporated in England by Royal Charter (Registered No. RC000524), registered office: Burlington House, Piccadilly, London W1J 0BA, UK, Telephone: +44 (0) 207 4378 6556.

Visit our website at www.rsc.org/books

Printed in the United Kingdom by CPI Group (UK) Ltd, Croydon, CR0 4YY, UK

Engineering Health
How Biotechnology Changed Medicine

Edited by

Lara V. Marks
University College London, UK
WhatIsBiotechnology.org
Email: l.marks@ucl.ac.uk

THE QUEEN'S AWARDS
FOR ENTERPRISE:
INTERNATIONAL TRADE
2013

Preface

The book grew out of the development of the website WhatIsBiotechnology.org. This is a non-profit educational resource that tells the stories of the people, places and tools behind biotechnology. The website aims to bring to life some of the challenges scientists face when translating laboratory research into diagnostic and therapeutic products and the impact that biotechnology has had, and is continuing to have, on the lives of patients and the working practices of clinicians. This is done through a rich display of materials collected from many different sources, including letters, laboratory notebooks, papers and photographs as well as interviews with key individuals.

Neither the website nor this book can fully capture all of the changes that biotechnology has made to medicine. They can only offer a snapshot of what is an ever-growing field. Yet, they provide a window into some of the exciting advances biotechnology has brought to medicine in recent years and their attendant risks and benefits.

This book could not have been written without the generosity of the many individuals who contributed to its different chapters. I am truly grateful to them all. Their hard work made my editorial work immeasurably easier. Many thanks also go to Rowan Frame and her team at the Royal Society of Chemistry for helping to guide the book to its completion.

Lara Marks

Engineering Health: How Biotechnology Changed Medicine
Edited by Lara V. Marks
© The Royal Society of Chemistry 2018
Published by the Royal Society of Chemistry, www.rsc.org

Contents

Chapter 1
Introduction: Biotechnology—An Ever Expanding Toolbox for Medicine
Lara V. Marks

Engineering Health: How Biotechnology Changed Medicine
Edited by Lara V. Marks
© The Royal Society of Chemistry 2018
Published by the Royal Society of Chemistry, www.rsc.org

Chapter 4
Monoclonal Antibodies: A Revolution in the Transformation of Healthcare 72
Lara V. Marks

Chapter 5
The Changing Fortune of Cancer Immunotherapy 97
Lara V. Marks

Chapter 6
Gene Therapy: An Evolving Story   **126**
Courtney Addison

Chapter 7
Stem Cells: An Emerging Field for Medicine   **147**
Alison Kraft and Frank Barry

CHAPTER 1

Introduction: Biotechnology—An Ever Expanding Toolbox for Medicine[†]

LARA V. MARKS

University College London, UK
Email: l.marks@ucl.ac.uk

1.1 INTRODUCTION

Biotechnology is intrinsic to medicine. Everywhere you look today, from medical research conducted in the laboratory through to the diagnosis and clinical treatment of a patient, biotechnology is pivotal to that process. Despite its importance, few non-scientists understand what biotechnology is, where it has come from or the many different functions it serves in everyday healthcare.

The term biotechnology was first coined in 1919 by Károly Ereky, a Hungarian agricultural engineer and economist. At its most basic level biotechnology refers to the controlled and

---

[†]Much of this chapter draws on material I originally wrote for the website WhatIsBiotechnology.org. I am grateful to Silvia Camporesi for comments on an earlier draft of this chapter.

Engineering Health: How Biotechnology Changed Medicine
Edited by Lara V. Marks
© The Royal Society of Chemistry 2018
Published by the Royal Society of Chemistry, www.rsc.org

deliberate manipulation of organisms and living cells to create products for the benefit of humans. In one form or another, humans have deployed biotechnology for thousands of years. Since prehistoric times, for example, they have used yeast to get bread dough to rise and to produce alcoholic drinks. Bacteria have also been added to milk for generations to make cheese and yoghurt. Animals and plants have also been selectively bred over many centuries to generate stronger and more productive off-spring for multiple purposes. In more recent years, increasing knowledge about how to manipulate and control the functions of various cells and organisms, including their genes, has given birth to a burgeoning number of products and technologies for combating human disease.

Biotechnology is currently one of the hottest growth areas in medicine. Between 2001 and 2012 investment in medical bio-technology research rose globally from £6.7bn to £66bn.[1] Such work is helping to determine the molecular causes of disease, to generate more accurate and faster diagnostic platforms and to develop drugs that are more precise in their target and person-alised for individual patients.

Just how important biotechnology has become can be seen from the fact that by 2013 seven out of ten of the best selling drugs were biological products. Also known as biologics, such drugs replicate natural substances in our body, including en-zymes, antibodies and hormones. They are made from a variety of natural resources—human, animal, and microorganism—and are usually manufactured using biotechnology techniques. Liv-ing entities, such as cells and tissues, also comprise biological products. Analysis in 2013 predicted that by 2018 biological drugs would account for a quarter of all drug spending world-wide and for more than 50 percent of the top selling drugs in the world. Such therapeutics include those that are manufactured in either animal cells or bacteria and make use of the body's natural immune system to fight disease.[2]

1.2   MOLECULAR DISEASE AND DNA

Much of the application of biotechnology in medicine is directed towards addressing structural changes at the molecular level that cause disease. This rests on the premise that an illness can be driven by an abnormality or deficiency of a particular molecule. Such thinking can be traced back to the work of Linus Pauling and colleagues at the California Institute of Technology in the late 1940s. Importantly, they demonstrated that sickle-cell anaemia, an inherited blood disorder, was linked to an abnormal haemoglobin, the protein responsible for delivering oxygen to cells in the body. They proposed the hypothesis that a mistake in the protein was caused by a defective gene. Pauling would go on to win the Nobel Prize in Chemistry for this work in 1954.[3] A gene is a distinct stretch of DNA (deoxyribonucleic acid) that carries the instructions needed to create proteins, specific molecules that are essential to the functioning of the body. Proteins not only do most of the work in cells they are also vital to the structure, function and regulation of the body's tissues and organs.

The concept that DNA could play a role in the disease process was highly novel for the time. DNA had been first discovered in the late nineteenth century, but remained little studied for many decades. In part this was due to the belief that DNA was an inert substance incapable of carrying genetic information because of its simple structure. Instead, proteins, which had a more complex structure, were assumed to act as genetic material. Attitudes to DNA began to shift as a result of some experiments by the physician and molecular biologist Oswald Avery and colleagues at the Rockefeller Institute in New York. In 1944 Avery showed that DNA could transform non-infectious bacteria associated with pneumonia into dangerous virulent forms.[4] Avery's work ignited a new interest in DNA. It would take time, however, for scientists to agree that it was DNA, not proteins, that carried genetic information. Consensus finally emerged after experiments conducted by the geneticists Alfred Hershey and Martha Chase at Cold Spring Harbor in 1952.[5]

By the 1950s a number of researchers had begun to investigate the structure of DNA in the hope that this would reveal how the molecule worked. The structure of DNA was finally cracked in 1953 as a result of the culmination of efforts by the biophysicists Rosalind Franklin, Maurice Wilkins and Ray Gosling, based at King's College London, and Francis Crick and James Watson based in the Cavendish Laboratory, Cambridge University. Their work showed DNA to be a long molecule made up of two strands coiled around each other in a spiral configuration called a double helix. Each strand was composed of four complementary nucleotides, chemical sub-units: adenine (A), cytosine (C), guanine (G) and thymine (T). The two strands were oriented in opposite directions so that adenines always joined thymines (A T) and cytosines were linked with guanines (C G). This structure helped each strand to reconstruct the other and facilitate the passing on of hereditary information.[6–8]

Soon after this breakthrough, in 1955, Fred Sanger, a biochemist at the William Dunn Institute of Biochemistry, Cambridge University, unveiled the molecular composition of the first protein: insulin. This protein, Sanger showed, had a specific sequence of building blocks, known as amino acids.[9] Sanger's finding was quickly seized upon by Crick, who by 1958 had developed a theory that the arrangement of nucleotides in DNA determined the sequence of amino acids in proteins and that this in turn regulated how a protein folded into its final shape.[10] Crick argued that it was this shape that decided each protein's function. He further proposed that an intermediary molecule helped the DNA to specify the sequence of the amino acids in a protein. The key question was how to prove his hypothesis.[11]

Crick recognised that one way to find out would be to investigate sickle-cell anaemia. Pauling and his colleagues had proposed the hypothesis that the difference in haemoglobin found in sickle-cell patients and healthy individuals could be down to a difference in the number of amino acids. How many amino acids were involved, remained unknown. Was it just one amino acid or more? Crick realised this could be resolved with the technique Sanger had developed to work out the composition of amino acids in insulin.

Based on this thinking, Crick launched a collaboration with Sanger and Vernon Ingram, a fellow colleague in the Cavendish laboratory. By 1957, after many hours of painstaking work, Ingram had determined that the difference between normal and sickle-cell haemoglobin was down to the replacement of 'only one of nearly 300 amino acids'.[12] Ingram's finding was a significant breakthrough. Not only did it challenge the scepticism of many scientists that the alteration of just one amino acid could produce a molecule as lethal as sickle-cell haemoglobin, it also marked the first time that anyone had managed to break the genetic code, the process by which cells translate information stored in DNA into proteins.[11] Ultimately the work on sickle-cell anaemia laid the foundation for a whole new approach in medicine, known as molecular medicine. Critically, it ignited a search for other genes or molecules that contributed to disease and ways to harness them for treatment.[13,14]

1.3   GENETIC ENGINEERING AND ITS CONTROVERSIAL BEGINNING

Unravelling the genetic process behind sickle-cell anaemia was just one investigation among many undertaken in the 1950s in order to understand the relationship between DNA and disease. Elsewhere microbiologists and biologists were examining the role of genetics in drug resistance. They were hunting for the biological mechanism bacteria use to resist viruses and other pathogens and thwart natural anti-microbial substances designed to kill or inhibit their growth. Such work was part of a broader effort to understand the mechanisms underlying the rising resistance of bacteria to antibiotic drugs, widely prescribed for medical treatment from the early 1940s. One of the fruits of such endeavours was the discovery of some biological mechanisms for manipulating and copying DNA.

Plasmids were one of the earliest biological tools scientists unearthed. Discovered in the 1940s, plasmids are small independent self-replicating strands of DNA that naturally exist in most bacteria and some fungi, protozoa, plants and animals. They come in a wide variety of lengths and provide the host organism with the necessary genes for coping with stress-related conditions, such as when encountering substances like antibiotics that impede their growth or threaten their survival. Plasmids have several useful characteristics. Firstly they contain only a small number of genes. Secondly, they snap quickly back into shape when cut open. Because of these features, scientists rapidly explored their use as a vehicle, or vector, for cloning, transferring and manipulating genes within the laboratory.

Soon after finding plasmids, scientists discovered some biochemical enzymes capable of cutting and pasting DNA. One of the first was polymerase, discovered in 1957. All living organisms make polymerase. It helps replicate a cell's DNA. Another important group of enzymes were restriction enzymes. This is a group of enzymes which bacteria use to cleave and destroy the DNA of invading viruses. Restriction enzymes were suggested to exist as early as 1952, but the first one was only isolated and characterised in the late 1960s. Often described as 'molecular scissors', restriction enzymes provided the means to cut DNA very precisely for the first time within the laboratory. By 1968 scientists had isolated another type of enzyme, known as ligase, which bacteria use to repair single-strand breaks in DNA. This provided an avenue for joining different DNA fragments together.

The discovery of plasmids and the different biochemical enzymes laid the foundation for the development of genetic engineering. This method involves selecting and cutting out a gene at specific point on a strand of DNA using restriction enzymes, and then inserting it into a plasmid to produce recombinant DNA. The very first piece of recombinant DNA was generated in June 1972 by Janet Mertz, a biochemistry graduate student working with Paul Berg at Stanford University, a subsequent Nobel Prize winner in 1980. This she did as part of a project to understand gene expression in human cells and its misregulation in cancer. Her recombinant DNA contained genes from the simian virus (SV40), a virus that lives in some monkey species, and a bacteriophage, a type of virus that infects bacteria.

Despite her achievement, Mertz was prevented from cloning the DNA. This involved inserting the recombinant DNA into bacteria for replication by its cell machinery. Mertz was unable to take the next step because of a controversy that broke out following her attendance at a workshop being run by Robert Pollack at Cold Spring Harbor Laboratory in June 1971. Pollack was alarmed to hear during the workshop that she was proposing to insert genes from SV40, into *Escherichia coli* (*E. coli*), bacteria that live in the guts of humans and other animals. SV40 is a largely harmless virus. While not known to cause any diseases in humans, SV40 had been shown within the laboratory to be capable of inducing the formation of tumours in rodents and human cells cultivated in culture. Pollack was particularly worried that some bacteria with the SV40 genes could escape from the laboratory, thereby infecting people and other mammals and possibly giving them cancer.

Mertz proposed the risk could be minimised by using an *E.coli* strain unable to survive outside of the laboratory. But Pollack continued to raise concerns. This persuaded Berg to self-impose a moratorium against anyone performing genetic engineering experiments in his laboratory that introduced SV40 genes into *E. coli* until the potential safety concerns had been addressed. In the end, the first cloning of recombinant DNA was carried out in June 1973 by Stanley Cohen and Herbert Boyer, based respectively at Stanford University and the University of California in San Francisco. They achieved this on the back of the methods originally outlined by Mertz.[15–17]

News of Boyer and Cohen's experiment immediately ignited a fierce public debate about the safety of genetic engineering. So great was the furore that in 1974 a group of American scientists agreed to self-impose a voluntary moratorium on experiments involving genetic engineering. This was lifted a year later following the introduction of strict guidelines drawn up by an international conference organised by Berg in Asilomar, California. The guidelines required laboratories to install tight security facilities to contain any experiments with recombinant DNA.[18] The Asilomar conference is now held up as a model of self-regulation by scientists. Indeed, it was recently used as a framework for experiments with CRISPR, a new form of gene editing described below. Significantly, Paul Berg played an important role in debating the guidelines for CRISPR.

Despite the initial controversy, genetic engineering soon grabbed the attention of venture capitalists, who swiftly began partnering with academic scientists to set up biotechnology companies to exploit the new technology. Genentech was the first company in the field, founded in 1976 in San Francisco. The new bioentrepeneurs envisaged inserting human genes into bacteria to encourage the production of unlimited quantities of heretofore scarce therapeutic proteins. Their vision was to spawn an entire new industry. The first two successful products prepared with genetic engineering were insulin, approved by the US Food and Drug Administration (FDA) in 1982, for the treatment of diabetes, and human growth hormone, approved by the FDA in 1985, to treat children with severely restricted growth. Since then genetic engineering has been used to generate medical drugs for a range of other diseases, including cancer, immune deficiency, HIV and heart attacks. As Chapter 2 by Alldread and Birch and Chapter 3 by Buckland show, the technique now underpins the production of many different drugs as well as vaccines. Yet, as these two chapters highlight, the adoption of genetic engineering in this sphere was not as easy or straightforward as originally anticipated. Indeed, it involved significant technical and scientific challenges on the manufacturing front.

1.4   MONOCLONAL ANTIBODIES

Within two years of Boyer and Cohen's successful cloning of recombinant DNA, another important tool appeared on the scene—a technique for the laboratory production of monoclonal antibodies (Mabs). These antibodies are derived from the millions of antibodies the immune system makes every day to fend off bacteria, viruses, pollen, fungi, and any other substance that can threaten the body, including toxins and chemicals. The first Mabs were produced in 1975 by Georges Köhler and Cesar Milstein, based at the Laboratory of Molecular Biology, Cambridge, UK. They developed this as part of their search for a research tool to investigate how the immune system produces so many different types of antibodies specifically targeted to the infinite number of foreign substances that invade the body.

While Köhler and Milstein were subsequently awarded the Nobel Prize in 1994, their invention attracted far less public fanfare at the time than the development of recombinant DNA. Over time, however, their innovation was to have an even more far-reaching impact in the medical field. Able to bind to specific markers found on the surface of cells, Mabs provided an important tool for the detection of unknown molecules and established their function for the first time. As Marks points out in Chapter 4, Mabs opened up new pathways for understanding multiple diseases on an unprecedented scale and greatly enhanced the speed and accuracy of diagnostics.[19] In addition they provided new avenues of treatment. Mab drugs are currently used to treat over 50 major diseases. Just how important they have become can be seen from the fact that Mab drugs now comprise six out of ten of the best-selling drugs worldwide and make up a third of all newly introduced medicines. Mabs are also now at the forefront of the development of immunotherapy for cancer, the subject of Chapter 5 by Marks. Hailed as one of the major advances in cancer treatment in recent years, such therapy is designed to induce, enhance or suppress the body's immune system to combat cancer.

1.5 DNA SEQUENCING

Just two years after Köhler and Milstein produced their first Mabs in the Laboratory of Molecular Biology, the same laboratory witnessed the development of another revolutionary technique: DNA sequencing. Devised by Fred Sanger in 1977, this procedure helps determine the exact order of the four building blocks, nucleotides, that make up a piece of DNA.[9,20] The method has four key steps. The first stage involves removing DNA from a cell. This can be done either mechanically or chemically. The second involves breaking up the DNA and inserting the fragments into cells that self-replicate so that they can be cloned. Once this is done, the DNA clones are mixed together with a dye-labelled primer (a short stretch of DNA that promotes replication) in a thermocycler, a machine that automatically raises and lowers the temperature to catalyse replication. Finally the DNA segments are placed on a gel and subjected to an electric current. Known as electrophoresis, this process helps sort out the DNA fragments by their size. When subjected to the electric current the smaller nucleotides move faster than the larger ones. The different nucleotide bases in the DNA are identified by their dyes, which are activated when they pass through a laser beam. All the information is fed into a computer and the DNA sequence is displayed on the screen for analysis (Figure 1.1).

Today DNA sequencing is one of the most important tools in medicine. What helped propel it forward was the development of another technique known as the polymerase chain reaction (PCR). Sometimes called 'molecular photocopying', PCR uses the enzyme polymerase and two matching strands of DNA to amplify and copy small segments of DNA. The founding principle for PCR was first developed in 1971 by Kjell Kleppe, a Norwegian researcher working in the laboratory of the Nobel Prize winning scientist Har Gobind Khorana at the Massachusetts Institute of Technology, but its practical application was only demonstrated in 1985, by the American biochemist Kary Mullis, who was awarded a Nobel Prize in 1993 for this work.[21]

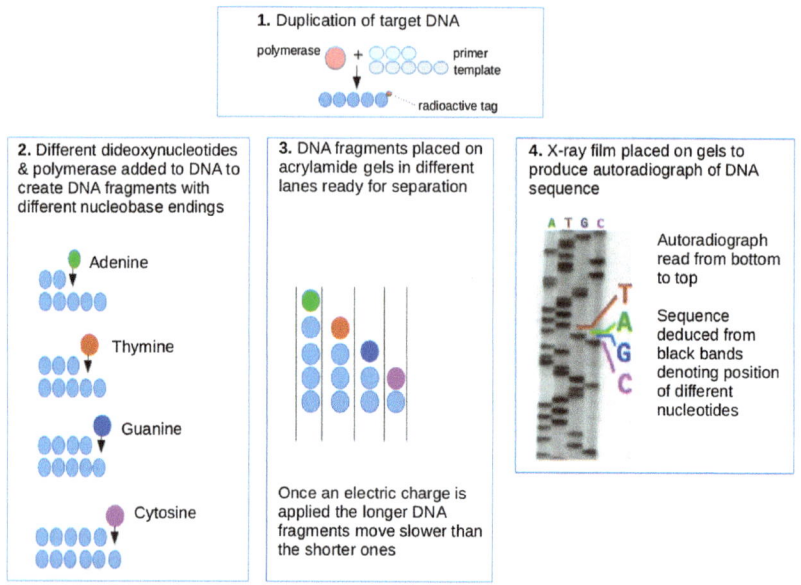

Figure 1.1   The basic steps in DNA sequencing.
Source: WhatIsBiotechnology.org (Reprinted with permission from Lara Marks).

PCR involves two basic steps. In the first instance a DNA sample is heated to separate its two strands. The two separated strands are then used as templates to synthesise two new DNA strands. This is done with the help of an enzyme called Taq polymerase. Once made the newly synthesised DNA strands are used as templates to generate two more copies of DNA. These steps are repeated multiple times with the help of a thermocycler. PCR can generate one billion exact copies of an original target of DNA within a couple of hours (Figure 1.2).

DNA sequencing is now an automated process. This has dramatically reduced both the time and cost of such work and the scale on which it can be carried out. Now it is possible to run DNA tests on thousands of different molecules in parallel. The technology has advanced so much that DNA sequencing has become a routine laboratory process. DNA can now be analysed from millions of samples of blood, semen and tissue from patients every year. Some idea of how far the technology has travelled can be seen from the fact that the first human genome

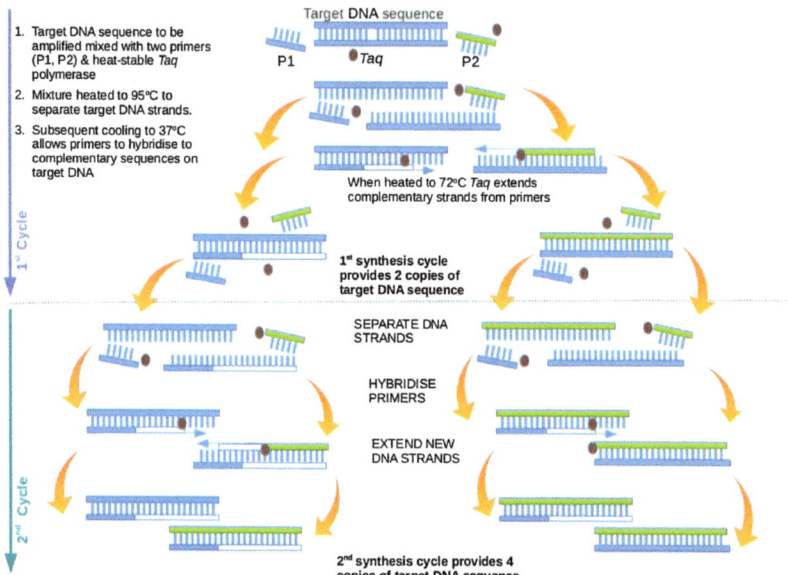

1. Target DNA sequence to be amplified mixed with two primers (P1, P2) & heat-stable *Taq* polymerase
2. Mixture heated to 95°C to separate target DNA strands.
3. Subsequent cooling to 37°C allows primers to hybridise to complementary sequences on target DNA

When heated to 72°C *Taq* extends complementary strands from primers

1st synthesis cycle provides 2 copies of target DNA sequence

SEPARATE DNA STRANDS

HYBRIDISE PRIMERS

EXTEND NEW DNA STRANDS

2nd synthesis cycle provides 4 copies of target DNA sequence

Figure 1.2   Diagram of PCR process.

sequenced by the Human Genome Project, started in 1990, cost nearly $3 billion and took 23 years to complete. By late 2015 it had become possible to sequence a whole human genome for just below $1500 and it could be performed in just 26 hours. Such whole-genome sequencing (WGS) is paving the way to greater understanding of the relationship between genetic variations and disease. New devices are currently being developed to help in the management of critically ill infants with suspected but undiagnosed genetic disease.[22,23]

Many of those who campaigned for the setting up of the Human Genome Project did so in the belief that it would pave the way to understanding the genetic code of life and open up a new chapter in medicine. It could take many more years, however, before researchers fully understand how instructions encoded in DNA shape health and disease. Such work will necessitate the investigation of genome samples taken from thousands, probably millions, of people. Nor will it be straightforward. Not only does the genetic profiling of individuals raise major ethical and legal concerns; many question the importance of genes in health compared with other factors such as environment and lifestyle. Often the genomic data does not reveal a neat relationship between genes and health.

1.6 WHOLE GENOME SEQUENCING—A WEAPON AGAINST ANTIMICROBIAL RESISTANCE

One area where WGS is already having a major impact is in the fight to control drug-resistant bacteria. These strains are a growing threat to public health globally. Traditionally most bacterial pathogens have been identified by growing a patient's specimen in a culture, testing its susceptibility to antimicrobial drugs and comparing it with other bacterial strains. The whole process is labour-intensive and time-consuming. Identification and susceptibility testing can take days, in the case of rapidly growing bacteria such as *E. coli*, or months, in the case of slower growing bacteria such as *Mycobacterium tuberculosis*.[22,23]

The exciting potential WGS technology holds can be seen from the case of the Rosie Hospital in Cambridge, UK. In 2011 three babies in the hospital's baby unit all tested positive at the same time for Methicillin-resistant *Staphylococcus aureus* (MRSA). On investigating the matter further the hospital's infection-control team discovered that a number of other babies had been infected sporadically by MRSA—sometimes months apart—over a period of 6 months. The team could not work out whether they were seeing an outbreak of MRSA or merely unrelated infections. Unable to answer the question using conventional methods the team called WGS experts from the Wellcome Trust Sanger Institute, Cambridge. The Sanger team quickly established that the samples were all related at the genome level, suggesting an outbreak. By sequencing other bacteria in the hospital's microbiology laboratory database with the same antibiotic susceptibility profile, they confirmed that this was a new strain of MRSA with twice as many infections as first thought and that it was also prevalent in the wider community. Two months after the previous infections were discovered, another baby tested positive despite the unit having just been thoroughly cleaned. So the team sequenced bacteria from swabs taken from staff on the unit and discovered that a particular individual was carrying the same strain of MRSA. Once that staff member was treated, the outbreak ended.[9]

The work at the Rosie Hospital was one of the first occasions when WGS was used to track down and halt the outbreak of an antimicrobial-resistant infection in a hospital. One of the advantages of the method was that it made it possible to reconstruct the history of the outbreak and identify its transmission route. The sequencing not only showed that the outbreak resulted from a new strain of MRSA, but also managed to link it to earlier infections on the ward and to trace its complex transmission pathways from babies to their mothers, from these mothers to other mothers in the postnatal ward, and to partners of affected mothers. What the sequencing research also highlighted was the pitfall of concentrating merely on hospital-based infection control, which pointed to the need to develop strategies that could take on board the important transmission dynamics between the community and healthcare institutions.[24]

WGS is not only being used to thwart the spread of antimicrobial-resistant infections in hospitals, but also in the community. British researchers are now pioneering a low-cost WGS tool for helping with tuberculosis, a growing public health threat. Usually it takes up to a month to confirm a diagnosis of TB, confirm treatment choices and to detect transmission between cases. This is major hindrance in terms of preventing the spread of the disease. Used directly on patient samples in the clinic, the new tool is being designed to provide results within the same day. Such a diagnostic would be a major breakthrough because it would make it easier to match patients to the right medication at the start of their treatment, thereby shortening the time they are infectious and thwarting the spread of drug-resistant tuberculosis.[25,26]

1.7 CRISPR: A NEW GENE EDITING TOOL AND CONTROVERSY

Today biotechnology stands on the cusp of another revolution as a result of the recent development of another method known as 'clustered regularly interspaced short palindromic repeats' or CRISPR for short. Like older genetic engineering techniques, CRISPR deploys the cellular machinery bacteria use to recognise and edit the DNA of harmful viruses. Where it differs is that it facilitates the introduction or removal of more than one gene at a time, cutting the process down from a number of years to a matter of weeks. This has made it possible to carry out genetic engineering on an unprecedented scale and at very low cost. Not being species-specific CRISPR can also be applied to organisms previously resistant to genetic engineering.

As was the case with recombinant DNA in the 1970s, the development of CRISPR has fuelled intense controversy. Where CRISPR is prompting most concern is in its use for editing the genomes of human embryos. Some of the first experiments carried out in this area were reported by Chinese scientists in April 2015. They used the technology to modify a gene responsible for the genetically inherited disease, beta-thalassaemia, a potentially fatal blood disorder. The editing was done in 86 human embryos deemed unviable for *in vitro* fertilisation because they were tripronuclear. Such embryos start off dividing like normal embryos but then stop developing because they have an abnormal collection of genes. Disappointingly only four of the 54 embryos tested had the desired genetic change. What was worrying was that many of the embryos displayed potentially harmful off-target mutations. Based on the poor performance of CRISPR the Chinese researchers cautioned against its use for editing viable human embryos.[27]

News of the Chinese experiment has triggered a major bio-ethical debate about how far CRISPR should be used.[28] The technology faces two major issues. The first is a philosophical dilemma. It centres on the extent to which CRISPR should be used to alter 'germ-line' cells—eggs and sperm—which are responsible for passing genes on to the next generation. While it will take many more years before the technology will be viable for creating designer babies, a public debate has already begun on this issue. So great was the fear that some scientists, including some who helped pioneer CRISPR, in December 2015 called for a moratorium on its use in germ-line cells at an international meeting organised in Washington DC by the US National Academy of Science (NAS) and the National Association of Medicine (NAM) together with the UK Royal Academy of Sciences and Chinese scientists. This recommendation was subsequently reversed by a report put out by NAS and NAM in February 2017 which instead of calling for banning such experiments argued for proceeding with caution.[29]

The second issue is one of safety. One of the major problems is that the technology is still in its infancy and knowledge about the genome remains very limited. Many scientists caution that the technology still needs a lot of work to increase its accuracy and make sure that changes made in one part of the genome do not introduce changes elsewhere which could have unforeseen consequences. This is a particularly important issue when it comes to the use of the technology for applications directed towards human health. Another critical issue is that once an organism, such as a plant or insect, is modified they are difficult to distinguish from the wild-type and once released into the environment could endanger biodiversity.

1.8 TRANSGENIC ANIMALS

Where CRISPR is already proving useful, however, is for the generation of transgenic animals. These are animals that have had their genomes altered by the transfer of genes or a gene from another species or breed. The first such animal, a mouse, was generated in 1974. Transgenic animals serve many different purposes in medicine. For example, they are routinely used as research models to examine the role of genes in terms of human disease susceptibility and progression as well as to test therapies. Animals are also genetically modified to produce complex human proteins, including Mabs, for the treatment of human disease.

In addition to helping in the production of protein therapeutics, transgenic animals are being explored as a means of growing human organs, which are in great shortage for transplants. A key pioneer in this field is George Church who with a team at Harvard University is harnessing CRISPR to develop transgenic pigs for this purpose. One of the key problems in transplanting organs from animals into humans is that they will be rejected by the human immune system. In 2015 Church's group reported that they had successfully used CRISPR to modify genes encoding molecules that sit on the surface of pig cells known to trigger human immune responses or cause blood clotting. They had also managed to inactivate retroviruses naturally embedded in the pig's genome. This is important because up to now organ transplants from pigs have carried the risk of transferring viruses into humans. The breakthrough by Church's team is highly significant because, while experimentally it has been shown that pig kidneys can function in baboons for months, safety concerns about viruses have prevented testing them out in humans.[30]

In the past it could take several years to engineer a transgenic animal. The process relied on genetically altering an animal's embryonic stem cells, injecting them into an embryo, and breeding multiple generations of animals. CRISPR has greatly reduced the number of steps in the process. Critically, it enables targeted genetic surgery of a fertilised egg. This has radically shortened the time needed to make a transgenic animal. Today, for example, it is possible to create a transgenic mouse in just 6 months.[31] Another advantage of CRISPR is that it makes it possible to alter more than one gene at a time. In October 2015 Church reported that he and his group had managed to modify more than 60 genes in pig embryos. This was ten times more than had been edited in any other animal (Figure 1.3).[32]

1.9   GENE THERAPY

In addition to helping create transgenic animals, CRISPR promises to aid gene therapy, an experimental technique that has been evolving since the 1980s. Such treatment uses a number of strategies, including the replacement of a mutated gene that causes a disease with a healthy copy of the gene, inactivating or 'knocking out' a mutated gene that functions improperly, or introducing a new gene into the body to combat a disease. What progress is being made in this field is explored by Addison in Chapter 6. As Addison points out, gene therapy

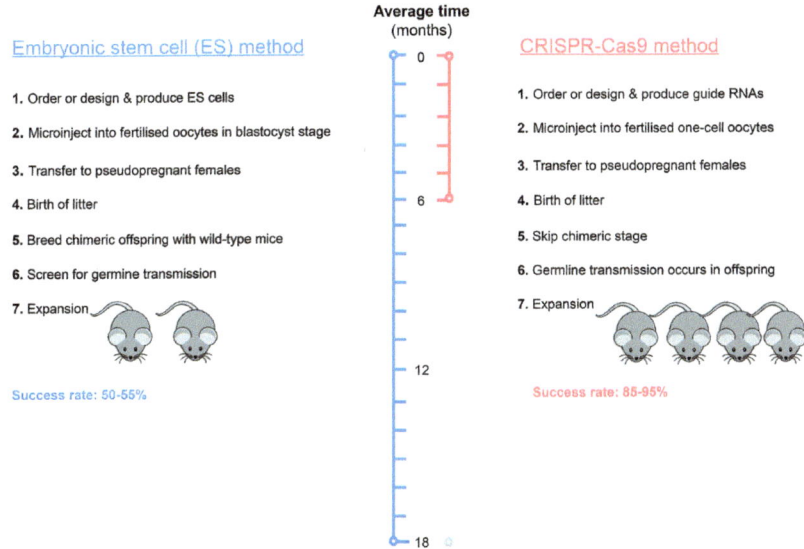

Figure 1.3   Time to produce transgenic mice with and without CRISPR.

has been dogged by controversy and its advance has been closely tied to debates about how far genetic intervention should go. This has not been helped by the fact that the very first patient treated with gene therapy, in 1990, died and that subsequent patients developed leukaemia. Added to this, there have been significant technical challenges in getting the therapy to work in humans.

Despite these set-backs gene therapy is now gaining renewed enthusiasm as a result of the development of CRISPR. Just how powerful CRISPR could prove for gene therapy is highlighted by the case of French teenager who in March 2017 was reported to have remained free of sickle-cell anaemia for 15 months following receipt of gene therapy. The procedure involved removing the patient's bone marrow, the part of the body that makes blood, and then genetically altering it in the laboratory using CRISPR. The aim was to modify it so that it no longer carried the DNA defect responsible for sickle-cell anaemia. Once done the bone marrow was infused back into the patient. The procedure, however, is still in its infancy and very expensive, which limits how far it can be rolled out to other patients.[33]

1.10   STEM CELLS

Gene therapy is not the only area to have experienced many ups and downs. So too has stem cell therapy, a field examined by Kraft and Barry in Chapter 7. The treatment takes advantage of the fact that stem cells are the body's master cells and have the ability to grow into any one of the body's more than 200 cell types. Kraft and Barry show that the term 'stem cell' was first used in the 1860s to describe the starting point in cell formation. However, it took many years to unravel the precise function of stem cells and for them to be deployed for therapy. This emerged on the back of the development of bone marrow transplants, a procedure introduced as a treatment for cancer and genetic blood disorders in the 1960s.

Many different types of stem cells have now been identified. The first, known as embryonic stem cells, are sourced from embryos formed during the blastocyst phase of embryonic development, which is four or five days after fertilisation. They are usually taken from human embryos left-over from *in vitro* fertilisation. The second, known as adult or mesenchymal stem cells, are found in different types of tissue, including bone marrow, blood, blood vessels, skeletal muscles, skin and liver. Stem cells can also be sourced from umbilical cord blood.

Like gene therapy, stem cell therapy has been surrounded by controversy. Most of this has centred on the use of human embryonic stem cells, which has prompted fierce ethical debates and strict regulatory oversight. Major advances, however, have been made in stem cell science in recent years. A critical breakthrough occurred in 2006. That year a Japanese team led by Kazutoshi Takahashi and Shinya Yamanaka discovered a way of genetically modifying somatic cells to reprogram them with properties similar to those of embryonic stem cells. The new cells are known as induced pluripotent stem cells (iPS). Such cells are now the subject of intense clinical investigation. Yamanka was awarded a Nobel Prize in 2012.

1.11   IMPACT IN THE CLINIC—EYE DISEASE

Where the new iPS cells appear to be most promising is in the treatment of blinding eye diseases, a topic discussed in detail in

Chapter 8 by Awwad, Khaw and Brochini. This chapter shows that stem cells are just one of a number of biological products now used to treat blinding eye diseases. Indeed, biotechnology is now paving the way to major breakthroughs for ophthalmic disorders. Some of the most notable successes have been with antibody-based drugs. Yet, as Awwad, Khaw and Brochini point out, the development of such treatments has not been straight-forward. Their chapter is a salutary reminder of just how complex the body's physiology is when it comes to applying a biological drug at the clinical level. As we can see from their chapter the physical and biochemical barriers of the eye make it a particularly challenging environment in which to administer treatment.

1.12 CELLULAR THERAPY: A NEW DISRUPTIVE TECHNOLOGY

Both stem cell and gene therapy are part of a wider group of products known as cellular therapies. This is a diverse spectrum of therapies, which use many different types of cells. These can be divided into two main categories: a patient's own cells (autologous therapies) and donor cells (allogeneic therapies). In many cases the therapy involves the development of bespoke cells for individual patients. The key aim of cellular therapy is either to replace a missing cell type, as in those that have a defective gene, the approach used for treating the French teenager with sickle cell anaemia described above, or to provide a necessary factor to boost the efficacy of a treatment, such as adoptive cellular therapy, described in Chapter 5, which enhances the power of T lymphocytes, white blood cells, for fighting cancer.

The field of cellular therapy has grown rapidly in recent years and can be considered the most recent phase of the biotechnology revolution in medicine. Most cellular therapies are still experimental and at a very early stage of clinical testing. The bulk of these have been directed towards rare disorders. While each of these affects only a small group of patients, when added up they represent a large population base. In 2009 it was estimated that the potential market for all cell-based therapies in the USA could exceed 100 million patients.[34,35]

Cellular therapies could offer significant benefits to patients who currently have limited or no treatment options. Yet, rolling them out on a wide-scale is significantly more complex than conventional small molecules or newer biological drugs. Unlike these therapeutic products, which are usually manufactured in a centralised industrial setting on a large scale, cellular therapies are largely bespoke treatments tailored to individual patients. This is a time-consuming process as it often involves removing cells from the patient, re-engineering them in the laboratory and then transfusing them back into the patient. In addition, it is expensive as the whole process must be conducted in a sterile environment that conforms with good manufacturing practice, all of which is difficult to maintain and run. On top of this the cells have short-half lives, meaning they only survive a few hours. Most cellular therapies therefore need to be produced close to the clinical setting of patients, so require multiple manufacturing sites. Another problem in terms of the commercialisation of cellular therapy is the diversity of cell types used in cellular therapy. This makes it difficult to develop a one-size-fits all manufacturing platform.[36] All of the issues outlined above have major implications in terms of cost. For example, the price of CAR-T therapy, an immunotherapy for cancer explored in Chapter 5, which an FDA panel recommended for marketing approval in July 2017, is predicted to be between $300 000 and $500 000 per treatment.[37]

1.13 COST IMPLICATIONS

While cellular therapies are likely to be the most expensive products on the market, they are part of a growing trend. Many of the newer biological drugs now available are much more expensive than conventional drugs. This, in part, reflects the fact that they tend to be derived from living material and involve more complex manufacturing processes. Furthermore, there is little competition in the marketplace to drive down the cost.

With many of the treatments promising to improve disease conditions heretofore difficult to treat, demand for them is rising. This is posing an impossible conundrum for policymakers wrestling with constrained healthcare resources. Nor is the price of the new drugs likely to be lowered in the future by the development of generic versions. This is because, unlike conventional drugs, biological drugs are difficult to replicate and require extensive clinical testing. Any inadvertent chemical modifications can affect their performance and safety. As a result, regulatory authorities require clinical trials to demonstrate equivalence for efficacy. This is not the case with other generics. To have sufficient power to be capable of detecting differences from the original drug, the clinical trial population often has to be very large. The biosimilar Mab, infliximab, approved in Korea in 2012 and in Europe in 2013, for example, underwent clinical testing in 874 patients in 20 countries across 115 sites. In addition to the costs of clinical testing, producing biosimilar Mabs requires state-of-the-art manufacturing technology, which is both expensive and cumbersome. Overall, analysts expect the complexities involved in manufacturing and testing biosimilar Mabs will keep costs high and that their prices will only be between 20 and 30% less than those of their patented counterparts. This is much lower than conventional drugs which cost between 80 and 90% less once off patent.[38]

1.14   SYNTHETIC BIOLOGY

One route to making more affordable drugs might lie in the emergence of synthetic biology examined by Race in Chapter 9 and Fletcher and Rosser in Chapter 10. This is an emerging discipline in biotechnology that seeks to combine the principles of engineering with the new knowledge about genetics and cell biology. Race's chapter provides an overview of some of the history behind the technology, showing how it has evolved since 1912, when the term was first coined, to its recent upsurge, which was aided by the development of recombinant DNA and the sharp fall in the cost of DNA sequencing. What is exciting is that it provides a means to create cells and tissues from scratch. As Race puts it, synthetic biology is conceptually like building a structure with LEGO® bricks.

So far most of the early applications of synthetic biology have been confined to relatively simple modifications of microbial cellular pathways to enhance their productivity for synthesizing certain medicines. The most well known example is that of artemisinin, an anti-malarial drug originally extracted from the plant sweet wormwood, which can now be produced using engineered yeast strains. This is particularly important because sweet wormwood harvests are fickle. By comparison, the application of synthetic biology in mammalian cells for the production of protein therapeutics is still in its infancy, as shown by Fletcher and Rosser in their chapter. Nonetheless, as they point out, the technology is already proving helpful in the production of vaccines, as demonstrated by the development of a synthetic flu vaccine in 2013.

1.15   CONCLUSION

As can be seen from this chapter, biotechnology is a fast and growing field in medicine. A book of this kind can therefore only scratch the surface of the many developments now happening. What is striking throughout is the complex and dynamic interface between different biotechnological tools in this process. Drug discovery and development, for example, makes use of a large range of biotechnology platforms simultaneously. It is highly dependent on DNA sequencing, PCR, monoclonal antibodies and genetic engineering. The same is true in the field of diagnostics and other medical applications.

What is often forgotten in this story is the increasingly important role of immunology in the development of biotechnology. Yet, as this chapter shows, many of the biotechnological tools we now have at our disposal were born out of a desire to understand the mechanisms of the immune system. Two of the key breakthroughs in biotechnology, genetic engineering and monoclonal antibodies, on which the biotechnology industry was built, grew out of such research. The tools for recombinant DNA, for example, originated from research into understanding how bacteria resist foreign invaders. Similarly, monoclonal antibodies emerged out of a search for a means to investigate the diversity of the immune system. In addition, many of the advances now being made in the field of cancer rest just as much on harnessing the immune system as on understanding the genetic predisposition of patients to the disease.

As will become apparent from reading the following chapters the advances now being made in medicine through the use of biotechnology are controversial. Furthermore, they are not inevitable and involve many blind alleys as well as successes. Many of the changes brought about by biotechnology in medicine have been incremental and hidden from view. Concentrating merely on what impact biotechnology has had on the development of drugs, which have often stolen the media limelight, misses the enormous changes it has brought to our understanding about the pathways of disease and the new tools now at our disposal for diagnosis.

REFERENCES

1. HM Government, Strength and opportunity 2013, *Annual Update*, 2013, https://www.gov.uk/government/uploads/system/uploads/attachment_data/file/298819/bis-14-p90-strength-opportunity-2013.pdf.
2. M. D. Dalzell, In 5 years, >50% of top-selling drugs will be biologics, *Managed Care* (October 2013) http://www.managedcaremag.com/archives/2013/10/5-years-50-top-selling-drugs-will-be-biologics. Accessed 18 Jan 2016.
3. W. A. Eaton, *Biophys. Chem.*, 2002, **100**(1–3), 109–116.
4. O. T. Avery, C. M. Macleod and M. McCarty, *J. Exp. Med.*, 1944, **79**, 137–157.
5. A. Hershey and M. Chase, *J. Gen. Physiol.*, 1952, **36**(1), 39–56.
6. J. D. Watson and F. H. C. Crick, *Nature*, 1953, **171**, 737–738.
7. M. H. Wilkins, A. R. Stokes and H. R. Wilson, *Nature*, 1953, **171**, 738–740.
8. R. Franklin and R. G. Gosling, *Nature*, 1953, **1717**, 740–741.
9. L. Marks, The Path to DNA Sequencing: The Life and Work of Fred Sanger, 2015, http://www.whatisbiotechnology.org/exhibitions/sanger. Accessed Apr 2016.
10. F. Crick, On protein synthesis, in *The Biological Replication of Macromolecules. Symposia of the Society of Experimental Biology*, ed. F. K. Sanders, Cambridge, 1958, vol. XII, pp. 138–163.

11. S. de Chadarevian, *J. Hist. Biol.*, 1996, **29**, 361–386.
12. V. M. Ingram, *Nature*, 1957, **180**, 326–328.
13. B. J. Strasser, *Am. J. Med. Genet., Part C*, 2002, **115**, 83–93.
14. B. J. Strasser, Biomedicine: Meanings, assumptions, and possible futures, *SSCI Report*, **1**, 2014, 1–44, http://biologie.unige.ch/fr/wp-content/uploads/sites/2/2012/10/SWIR-1-2014_Biomedicine_140327.pdf. Accessed 5 Apr 2016.
15. L. V. Marks, Janet Mertz, http://www.whatisbiotechnology.org/people/Mertz. Accessed Mar 2017.
16. D. Yi, *The Recombinant University*, Chicago, 2015, pp. 87–99.
17. E. C. Friedberg, *A Biography of Paul Berg: The Recombinant DNA Controversy revisited*, Hakensack, NJ, 2014.
18. D. S. Fredrickson, Asimolar and Recombinant DNA: The end of the beginning, in *Biomedical Politics*, ed. K. E. Hanna, Washington DC, 1991, pp. 258–298.
19. J. L. Marx, Monoclonal antibodies and their applications, in *A Revolution in Biotechnology*, ed. J. L. Marx, Cambridge, 1989, pp. 145–159, 147–148.
20. M. Garcia-Sancho, *Biology, Computing and the History of Molecular Sequencing: From Proteins to DNA, 1945-2000*, Basingstoke, 2012.
21. P. Rabinow, *Making PCR: A Story of Biotechnology*, Chicago, 1996.
22. N. A. Miller, E. G. Farrow, M. Gibson, L. K. Willig, G. Twist, B. Yoo, T. Marrs, S. Corder, L. Krivohlavek, A. Walter, J. E. Petrikin, C. J. Saunders, I. Thiffault, S. E. Soden, L. D. Smith, D. L. Dinwiddie, S. Herd, J. A. Cakici, S. Catreux, M. Ruehle and S. F. Kingsmore, *Genome Med.*, 2015, 7, 100.
23. NIH, *The Cost of Sequencing a Human Genome*, July 2016, https://www.genome.gov/sequencingcosts/. Accessed Mar 2017.
24. S. R. Harris, E. J. P. Cartwright, M. E. Török, M. T. G. Holden, N. M. Brown, A. L. Ogilvy-Stuart, M. J. Ellington, M. A. Quail, S. D. Bentley, J. Parkhill and S. J. Peacock, *Lancet Infect. Dis.*, 2013, **13**(2), 130–136.
25. Public Health England, England world leaders in the use of whole genome sequencing to diagnose TB, 28 March 2017, https://www.gov.uk/government/news/england-world-

leaders-in-the-use-of-whole-genome-sequencing-to-diagnose-tb. Accessed Mar 2017.

26. A. A. Votintseva, P. Bradley, L. Pankhurst, C. del Ojo Elias, M. Loose, K. Nilgiriwala, A. Chatterjee, E. G. Smith, N. Sanderson, T. M. Walker, M. R. Morgan, D. H. Wyllie, A. S. Walker, T. E. A. Peto, D. W. Crook and Z. Iqbal, *J. Clin. Microbiol.*, 2017, **55**(4), 1–48.

27. P. Liang, Y. Xu, X. Zhang, C. Ding, R. Huang, Z. Zhang, J. Lv, X. Xie, Y. Chen, Y. Li, Y. Sun, Y. Bai, Z. Songyang, W. Ma, C. Zhou and J. Huang, *Protein Cell*, 2015, **6**(5), 363–372.

28. S. Camporesi and G. Cavaliere, *Pers. Med.*, 2016, **13**(6), 575–586.

29. NAS and NAM, *Human Genome Editing: Science, Ethics and Governance*, February 2017, https://www.nap.edu/read/24623/chapter/1. Accessed Apr 2017.

30. Wyss Institute, Removing 62 Barrier to Pig-to-human Organ Transplant in One Fell Swoop, 11 October 2015, https://wyss.harvard.edu/removing-62-barriers-to-pig-to-human-organ-transplant-in-one-fell-swoop/. Accessed Apr 2017.

31. J. Cohen, Any idiot can do it. Genome editor CRISPR could put mutant mice in everyone's reach, *Science Magazine*, 3 November 2016, http://www.sciencemag.org/news/2016/11/any-idiot-can-do-it-genome-editor-crispr-could-put-mutant-mice-everyones-reach. Accessed Mar 2017.

32. S. Reardon, Gene editing record smashed in pigs, *Nature*, 2015, DOI: 10.1038/nature.2015.18525.

33. J.-A. Ribell, S. Hacein-Bey-Abina, E. Payen, A. Magnani, M. Semeraro, E. Magrin, L. Caccavelli, B. Neven, P. Bourget, W. El Nemer, P. Bartolucci, L. Weber, H. Puy, J.-F. Meritet, D. Grevent, Y. Beuzard, S. Chrétien, T. Lefebvre, R. W. Ross, O. Negre, G. Veres, L. Sandler, S. Soni, M. de Montalembert, S. Blanche, P. Leboulch and M. Cavazzana, *N. Engl. J. Med.*, 2017, **376**, 848–855.

34. C. Mason, D. A. Brindley, E. J. Culme-Seymour and N. L. Davie, *Regener. Med.*, 2011, **6**(3), 265–272.

35. N. M. Mount, S. J. Ward, P. Kefala and J. Hylliner, *Philos. Trans. R. Soc. London, Ser. B*, 2015, **370**(1680), 1–16.

36. T. R. J. Heathman, A. W. Nienow, M. J. McCall, K. Coopman, B. Kara and C. J. Hewitt, *Regener. Med.*, 2015, **10**(1), 49–64.

37. A. Ward, Cancer therapy re-engineers cells to hunt and destroy, *Financial Times*, 2 June 2016.

38. P. Seymour, First monoclonal antibody submitted to EMA for biosimilar approval, Bioprocess blog (2 May 2012), http://www.bioprocessblog.com/archives/409. Accessed Mar 2017.

CHAPTER 2

Biopharmaceutical Proteins: The Manufacturing Challenge

RICHARD ALLDREAD*[a] AND JOHN BIRCH*[b]

[a] National Biologics Manufacturing Centre, UK; [b] University College London, UK
*Email: richard.alldread@uk-cpi.com; john.birch16@btinternet.com

2.1 INTRODUCTION

The biopharmaceutical industry is concerned with the production of protein molecules from living cells for use as therapeutic agents. Such proteins can be relatively simple, *e.g.* human growth hormone; or highly complex, *e.g.* a monoclonal antibody (Mab), and are produced in a variety of cell types from bacteria through to mammalian cells.

From relatively humble beginnings the biopharmaceutical industry has grown to become a multibillion dollar industry of global importance. Biopharmaceutical products underpin the businesses of many major pharmaceutical companies and increasingly drive their profits. Table 2.1, showing the global sales of prescription medicines in 2014, gives some idea of the current financial importance of biopharmaceutical products. Out of the top ten best-selling drugs that year, seven were

Engineering Health: How Biotechnology Changed Medicine
Edited by Lara V. Marks
© The Royal Society of Chemistry 2018
Published by the Royal Society of Chemistry, www.rsc.org

Table 2.1 Global sales of leading prescription medicines in 2014. Data taken from ref. 1.

Product	Company	Indication(s)	Molecule type	Sales (Mio USD)
Humira®	AbbVie	Autoimmune	Mab	13 021
Sovaldi/ Harvoni®	Gilead	Hepatitis C	Small	12 410
Remicade®	Johnson & Johnson/ Merck	Autoimmune	Mab	10 151
Enbrel®	Pfizer/Amgen	Autoimmune	Mab fusion	9120
Lantus®	Sanofi	Diabetes	Insulin	8152
Rituxan®	Roche	Leukaemia/ autoimmune	Mab	7356
Avastin®	Roche	Cancers	Mab	6841
Seretide/ Advair®	GSK	Asthma	Small	6700
Herceptin®	Roche	Breast cancer	Mab	6690
Crestor®	AstraZeneca	Cholesterol	Small	6617

biopharmaceuticals, six of these being Mab-related products. In addition to its financial importance, the industry is increasingly seen as important to meeting society's future healthcare needs. To achieve this goal, however, further developments will need to be made to the way biopharmaceuticals are developed and manufactured.

2.2 THE EMERGENCE OF A NEW TECHNOLOGY

In 1974 a group of American scientists published a paper demonstrating that DNA could be transferred from an animal (the African clawed frog) to a bacterium and was able to function in this simple organism.[2] To the non-specialist this may, perhaps, have seemed a little esoteric, but to those working in the field it highlighted how a new science known as genetic engineering, or recombinant DNA technology, could become an invaluable tool for both biological research and healthcare applications.

Figure 2.1 illustrates the basic steps involved in recombinant DNA technology. It has a range of components, including plasmids (normally short circular pieces of DNA that can replicate within cells), bacteriophages (essentially viruses that infect and replicate in bacteria), restriction endonucleases

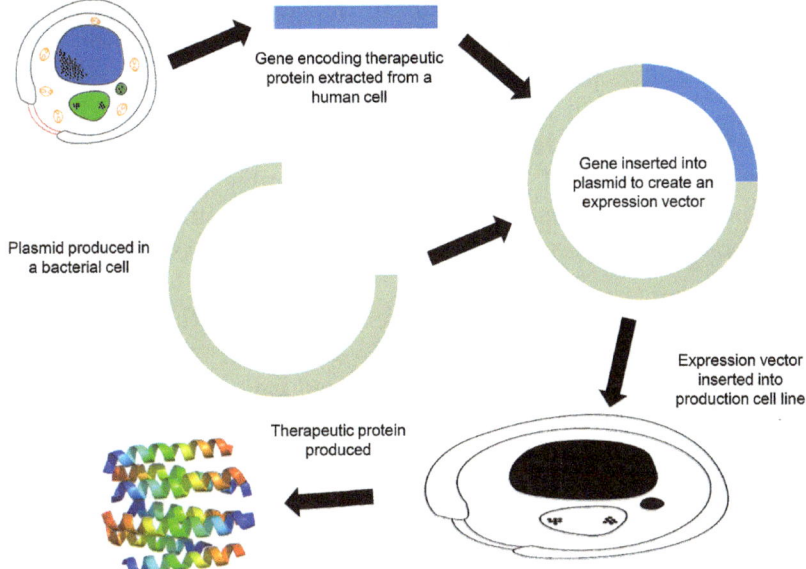

Figure 2.1   Principles of recombinant DNA technology.

(bacterial enzymes that recognise and cut a particular DNA sequence) and DNA ligase (an enzyme which can join DNA ends together).

Where genetic engineering immediately proved useful was in the field of therapeutic proteins. The clinical potential of human proteins had been recognised for many decades, and indeed a small number, such as insulin and human growth hormone, were already in use for treatment by the 1970s. However, these were exceptions; for the majority of interesting proteins there was simply no practical way of making them in sufficient quantity. Genetic engineering offered one means to solve this problem. Scientists soon began to explore whether they could transfer human genes for proteins of interest to bacteria, which could then be used as cell 'factories' to make the proteins on an industrial scale using the same type of fermentation technology already used to make antibiotics and industrial enzymes.

By 1979 Genentech, a company established in California to exploit the new genetic engineering technology, had successfully expressed genes for human insulin in a harmless version of the gut bacterium, *Escherichia coli* (*E. coli*).[3] Genentech soon partnered with the pharmaceutical company Eli Lilly to take the venture further. Eli Lilly had both a long history of making insulin from animal-derived material and a strong expertise in large-scale microbial fermentation technology to make other products. In 1982 insulin became the first recombinant (genetically engineered) therapeutic protein to be licensed for use in humans.[4,5]

The decision to deploy genetic engineering first for manufacturing insulin might seem odd given that very large quantities of the protein could already be produced from the pancreases of pigs, but concern was growing that, with the increasing incidence of diabetes, this source might not meet future needs.[6] In addition, pig insulin has a slightly different composition to the human version so can provoke an immune response in some patients.

By the end of the decade two other recombinant products had been successfully produced in *E. coli*. The first was human growth hormone. Human growth hormone, like insulin, was already available but its source, pituitary material from human corpses, was far from ideal. Not only did it provide a limited amount of the hormone, it ran the risk of being infected with Creutzfeldt–Jakob Disease (CJD), an infectious disease. Production in bacteria completely removed this risk.

The next product manufactured with the help of genetic engineering in *E. coli* was interferon, a type of protein first discovered in 1957 by Alick Isaacs and Jean Lindenmann. Interferons are a natural group of proteins the body produces as a part of its defence mechanism against pathogens. While the discovery of interferon sparked immediate excitement that it could provide a 'wonder drug' for combating viral diseases like hepatitis and some cancers, progress was hampered by the fact that the protein was a scarce commodity. Not only does the body produce only minute quantities of interferon but its laboratory production was severely limited, being reliant on human cells in culture being treated with viral agents to induce production of interferon.[7] Overall, it was a major challenge to make enough interferon for clinical use, even allowing for its exceptional potency, which means that only small doses are required. Genetic engineering was an elegant solution to the problem. It enabled different interferon types to be produced in bacterial culture, in some cases at titres of grams per litre, rather than the micrograms per litre previously seen.[8]

Over the ensuing decade many other products would be developed with the help of genetic engineering. Since the first approval of recombinant insulin in 1982 over 300 biopharmaceutical products have been approved worldwide and many more are currently in development. Table 2.2 lists some major milestones in the development of the industry. This has been marked by the development of increasingly complex products.

Table 2.2 Milestones in the development of the biopharmaceutical industry. Adapted from R. M. Alldread, J. R. Birch, H. K. Metcalfe, S. Farid, A. J. Racher, R. J. Young, M. Khan, Industrial scale suspension culture of living cells, Chapter 3.4 Large Scale Suspension Culture of Mammalian Cells. John Wiley & Sons, 2014 © 2014 Wiley-VCH Verlag GmbH & Co. KGaA, Boschstr. 12, 69469 Weinheim, Germany.[9]

Year	Product	Company	Cell line	Notes
1982	Insulin (Humulin®)	Eli Lilly	*E.coli*	First recombinant product
1983	Insulin (Novolin®)	Novo Nordisk	*S. cerevisiae*	First recombinant product made in a yeast
1987	Tissue plasminogen activator (Activase®)	Genentech	CHO	First recombinant product made in a mammalian cell
1989	Epoeitin (Epogen®/ Procrit®)	Amgen Johnson & Johnson	CHO	First mammalian-produced blockbuster product
1992	Factor VIII (Recombinate®)	Baxter	CHO	High molecular weight protein (over 200 kDa)
1994	Platelet Mab Fab (ReoPro®)	Centocor	NS0	Antibody-derived product
1997	Il-2 Receptor Mab (Zenapax®)	Roche, PDL Biopharma Biogen Idec	NS0	First humanised monoclonal antibody
1997	CD20 Mab (Rituxan®)	Genentech Biogen Idec	CHO	First monoclonal antibody from CHO cell line
2000	Gemtuzumab (Mylotarg®)	Wyeth	NS0	First monoclonal antibody conjugate
2002	Adalimumab (Humira®)	Abbott	CHO	First fully human monoclonal antibody

2.2.1   Other Host Organisms for Production

Many recombinant proteins were initially produced in *E. coli* or in the yeast *Saccharomyces cerevisiae*. These organisms were chosen for production because they were already well understood in the laboratory and could be grown readily on a large scale. It soon became apparent, however, that they were not suitable for all products of interest. Insulin and growth hormone are small and relatively simple proteins. As researchers turned their attention to more complex proteins it soon became clear that microbes simply did not have the machinery to handle the complexity of some human proteins.

Proteins are made up of long chains of amino acids, and in order to function these chains have to be folded in very particular ways. For some proteins this can be difficult to achieve in microorganisms. In addition, many human proteins are modified by the addition of other molecules, particularly sugars (a process known as glycosylation) which is an important determinant of their biological and therapeutic effects. Microbes either do not make these modifications or, as in the case of yeast, make modifications that are not in the desired human format.

The logical conclusion was to turn to mammalian cells that naturally make complex proteins. Many of the techniques developed for mammalian cells were similar to those developed for microbes. A crucial difference, however, was that in bacterial hosts the introduced gene was generally maintained on a self-replicating piece of DNA (plasmid) separate from the host chromosome, whereas in higher mammalian cell systems, the gene was normally integrated into the host genome and so maintained and replicated as part of a natural chromosome.

2.2.2   Mammalian Cell Culture

Although mammalian cell cultures did not have the long history of industrial application that underpinned production in microorganisms, they had been used for laboratory research purposes for many years. Their use grew out of the work of Ross Harrison, a scientist based at Johns Hopkins University, USA. He established in the early 1900s that nerve cells could be grown in culture fluids outside the body.[10] Following him, other

Table 2.3 Significant milestones in the development of mammalian cell culture. R. M. Alldread, J. R. Birch, H. K. Metcalfe, S. Farid, A. J. Racher, R. J. Young, M. Khan, Industrial scale suspension culture of living cells, Chapter 3.4 Large Scale Suspension Culture of Mammalian Cells. John Wiley & Sons, 2014 © 2014 Wiley-VCH Verlag GmbH & Co. KGaA, Boschstr. 12, 69469 Weinheim, Germany.[9]

Year	Event
1885	Maintenance of embryonic chick cells *in vitro* in a saline solution demonstrated
1913	Long term growth of cell cultures shown by feeding under aseptic conditions
1948	First permanent (immortal) cell line established, the L line (mouse subcutaneous tissue)
1951	First permanent human cell line established, HeLa, a cell line derived from a cervical cancer
1953	First suspension culture of mammalian cells
1957	Chinese Hamster Ovary (CHO) cell line established
1957	First growth of a cell line in stirred tank reactors
1960s	Development of large-scale suspension cultures of BHK (Baby Hamster Kidney) cells for Foot and Mouth disease vaccine production
1965	Development of fully defined (serum-free) medium for certain cell types
1975	First monoclonal-antibody-secreting hybridoma cell lines produced
1977	Efficient method for introducing single-copy mammalian genes into cultured cells developed

researchers developed many cultures of different mammalian cell types. Table 2.3 lists some of the key milestones in the development of mammalian cell culture. One of the key developments was the establishment of permanent cell lines, the first being a mouse-derived line in 1948. Whereas previously isolated cell lines could only be grown for a limited number of generations, those that are permanent are genetically modified to proliferate indefinitely, *i.e.* immortalised, and can be grown continuously in culture.[11]

Another important source were cell lines developed for vaccine production. This included monkey kidney cells. These were first utilised for the study of human virus pathogens and went on to become important to the development of now commonly used virus vaccines, the first of which was the Salk polio vaccine introduced in 1954. Subsequently, cells derived from human tissue were also used for vaccine production.

The types of cells that were historically used for vaccine production have two characteristics that were a major disadvantage for manufacturing proteins. They are only capable of a finite and limited number of cell divisions and they are "anchorage dependent"; *i.e.* they will only grow and multiply attached to a surface. This is not too much of a problem for many vaccines, where manufacturing is on a small scale and cells can be grown attached to the internal surfaces of glass or plastic flasks and bottles. Such methods, however, are unsuitable for manufacturing many biopharmaceutical proteins, where much larger amounts of the product are required, which would need vast numbers of bottles. For these proteins we ideally need cells that behave more like microorganisms *i.e.* having the capacity to multiply indefinitely and grow suspended in a culture medium in a container, known as a bioreactor, which can be scaled up by volume rather than by numbers.

One cell type that has proven invaluable to the biopharmaceutical industry is the Chinese Hamster Ovary (CHO) cell line. First isolated in 1957 CHO has become the cell of choice for the production of recombinant proteins, including biopharmaceuticals.[12–14] CHO cells have a number of characteristics that are attractive for the production of therapeutic proteins: they are robust, grow rapidly in culture, carry out modifications of proteins in a similar way to human cells, can have genetic material inserted into them easily and, because they are from a species sufficiently distant from humans, are relatively safe from the risk of cross transmission of harmful viruses. The CHO cell line was used by Genentech to make tissue plasminogen activator, a protein approved for the market in 1987, which dissolves blood clots. This was the first genetically engineered therapeutic protein to be produced in mammalian cell culture.

Descendants of the CHO cell line are now very widely used for the manufacture of a range of biopharmaceuticals, something that would have sounded like science fiction in the 1950s when the cell line was created. In fact more than half of all biopharmaceutical proteins are made in mammalian cells, mostly in CHO, although one or two other cell lines, mostly derived from mice and hamsters, are used for some products (Table 2.4). The descendants of the original CHO cell line have become quite a varied family. Over the years variants have been selected with

Table 2.4 Cell factories for making recombinant proteins. Data taken from ref. 15 and 16.

Organism		Approximate number of proteins (USA & EU)	Examples
Bacteria	*Escherichia coli*	33	Insulin, growth hormone, interferon
Yeasts	*Sacharomyces cerevisiae*	13	Insulin, hepatitis vaccine,
	Pichia pastoris	2	Plasma proteins
Mammalian cells	Chinese Hamster Ovary (CHO)	54	Antibodies, erythropoietin, enzymes
	NS0 (mouse)	10	Antibodies
	Sp2/0 (mouse)	9	Antibodies
	Other mammalian	9	Blood clotting factors, enzymes
Insect cells		2	Vaccines (flu & papilloma viruses)
Transgenic (genetically engineered)	Plant cells	1	Enzyme
	Animal milk	2	Enzyme inhibitors
	Chicken eggs	1	Enzyme

particularly desirable characteristics, for example the ability to grow spontaneously in suspension, avoiding the long period of adaptation required for the parent line. The CHO cell line is likely to remain the cell of choice for therapeutic protein production.

In recent years there has been an increasing use of genetic engineering to modify CHO cells. One particularly significant advance has been in the area of glycoengineering; modifying the sugar molecules that are added to proteins. An example of this involves modifying the level of the sugar fucose in antibodies, proteins the body makes to combat foreign invaders. Fucose is one of the sugars commonly found in antibodies and it has been found that its absence can enhance some aspects of biological activity which are desirable in some therapeutic applications. CHO cells have been engineered to knock out the enzyme which makes fucose, resulting in a cell line which can be used as a host to make fucose-free antibodies.[17]

What has also been critical to the development of mammalian cell culture for biopharmaceutical production has been the availability of improved techniques and materials for the culture of cell lines. Systematic studies of the nutritional requirements of mammalian cell lines in particular have helped the development of better media and systems to control the physicochemical environment of cultured cells. Another important landmark for the economics and safety of mammalian cell culture was the demonstration that certain mammalian cell types commonly used for manufacturing, including CHO, could be grown without the addition of animal-derived materials.[18] Today a vast array of commercial cell culture media is available for a variety of cell lines and uses. This has made cell culture a technique accessible to many researchers and suitable for commercial purposes.

Productivity is not only affected by the type of culture media used in production, but also by type of product being made, as some proteins are inherently more difficult to make. Another important influence is the structure of the genetic material used to express the protein. This is affected by the fact that the genetic code has built in redundancy, in other words there are different genetic sequences capable of coding for the same amino acid sequences in the protein. Some sequences work better than others, depending amongst other factors, on the type of organism used for production. Optimisation of the DNA sequence is now routine.

2.2.3 Alternative Production Systems

Whilst the vast majority of protein products are made in bacterial, yeast or mammalian cell systems there are a small number of exceptions where other host systems are used (Table 2.4). Insect cells, for example, have proven useful for the production of some recombinant viral protein vaccines, two of which are in use against flu and cervical cancer.[15,19] Another route that has been explored has been the use of genetically engineered, or transgenic, animals and plants to make proteins. Today four products are being made respectively in sheep's milk, rabbit's milk, hen's eggs and plant cells.[15,20] Yet, while transgenic systems have attracted a great deal of investment, progress has been slow in this area. Moreover, it is debatable whether transgenic systems will be widely adopted in the future given the extraordinary progress made, particularly with animal cell culture.

2.2.4   Generating Cells for the Production Process

Before manufacturing can begin, a host organism, whether it is a microorganism or mammalian cell, needs to be genetically modified to enable it to express the desired protein. Expression is the process by which genetic information is converted into instructions for making proteins. To achieve this, the product gene is inserted into a longer length of DNA, known as a vector, which includes other genetic elements which ensure efficient expression of the product gene. The vector also, usually, contains a gene which acts as a selectable marker, such as resistance to a toxic drug. This simplifies the process of finding that portion of the cell population that has incorporated the vector and with it the product gene. A range of chemical and physical methods are available for getting genes into cells.

Within the population of productive cells there will still be considerable variation in productivity between individual cells and the way populations derived from these cells grow will also vary under production conditions. Hence a screening process is carried out to isolate cells with the best growth and productivity characteristics. Many of the screening steps are now automated, including the isolation of single cells and the culture of populations derived from these "clones" in miniature bioreactors, which mimic the eventual production process.

Once a clonally derived cell line with the desired properties has been selected it has to be preserved. A product may be manufactured over many years and the production strain has to be available for the lifespan of the product. To address this 'cell banks' are established. A culture is grown, and the cells are separated and distributed into small containers known as vials, which are then frozen. In the case of mammalian cells this is typically in liquid nitrogen refrigerators at $-180\,^{\circ}\text{C}$. In this state of suspended animation, cells can survive for extremely long periods; certainly many years. The initial bank to be prepared is known as the master cell bank and cells from the bank are tested rigorously to establish their suitability, both in terms of manufacturing capability and safety. One needs to be assured that growth, productivity and product composition will be consistent over the number of generations needed to grow the cells to manufacturing scale. Tests are also carried out to ensure that the cell stocks are not harbouring contaminating microorganisms which could influence the process, and more importantly could potentially be hazardous. Secondary banks, called working cell banks are generated from the master cell bank and each manufacturing batch starts with a vial from a working bank. A vial is thawed and a culture is established, which is then expanded through a series of culture vessels of increasing volume until there are enough cells to seed the production reactor.

2.3   MONOCLONAL ANTIBODY TECHNOLOGY

In addition to recombinant DNA technology helping to scale up the production of almost any protein of therapeutic interest, the biopharmaceutical industry was revolutionised by another technique known as hybridoma or monoclonal antibody technology. This technique generates antibodies, large Y-shaped proteins normally produced by the immune system to help identify and remove, for example, infectious agents and cancer cells. A very large number of antibodies can be generated, each able to bind to a specific site, called an antigen, on the target molecule or organism.

The hybridoma method was first published in 1975 by Georges Köhler and César Milstein.[21] Figure 2.2 shows the basic method, a live mouse is challenged with the target (an antigen) causing it to mount an immune response directed at the target, immune cells are isolated from the mouse and fused with a type of tumour cell to produce a hybridoma, a hybrid cell that has the ability to grow continuously (the property of the tumour cell) whilst producing an antibody (the property of the immune cell). Each antibody is identical, giving them the name monoclonal antibodies (Mabs), and targets a specific antigen. Once made the hybridoma cell producing the desired antibody can be selected, cloned and grown in unlimited amounts.

Initially the hybridoma technology could only produce mouse antibodies, which have limited applications as human therapeutics. Nevertheless, mouse antibodies quickly proved to be useful reagents for research and for a wide variety of tests for measuring biological substances, such as hormones, in bodily fluids. One of the first clinical applications for which Mabs were

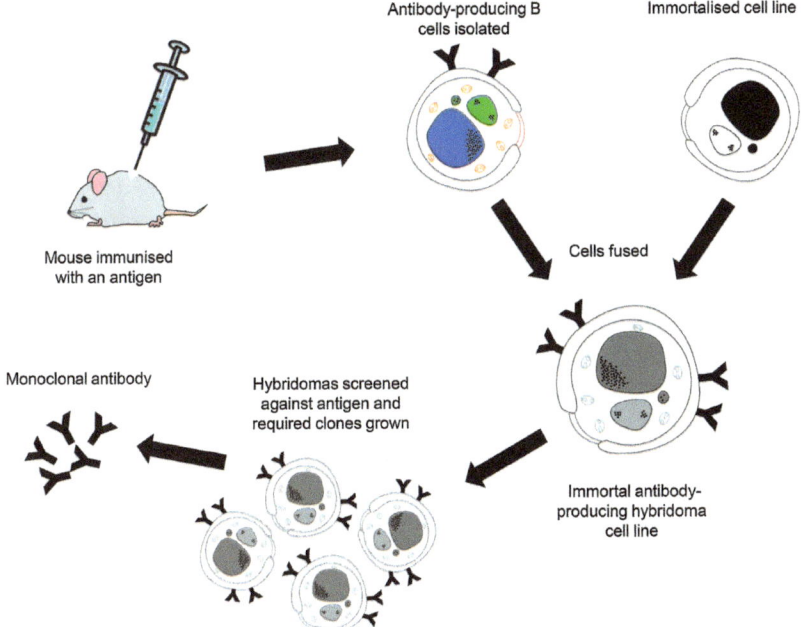

Figure 2.2   Monoclonal antibody process.

deployed was in tests for typing blood, being used as a substitute for reagents originally generated from human blood. This posed a significant challenge in terms of production, requiring many kilograms of Mabs per year. By 1986 Celltech, a UK biotechnology company had managed to meet this demand by growing the hybridoma cells in 1000 litre stainless steel bioreactors. Antibody titres were fourfold to fivefold higher than in laboratory culture systems as a result of optimisation of the culture medium and process conditions in the reactor and optimising the culture. Celltech's system was one of the first large-scale cell culture processes for antibody production.[22,23]

One of the major problems with the early hybridoma technology was that the resulting antibody was derived from a mouse, so had the potential to prompt an immune response if administered to humans. This prevented their adoption as therapeutics. By the late 1980s, however, scientists had resolved the issue by genetically engineering mouse antibodies to have more human-like characteristics which opened the way to their use as drugs. The 'humanisation' of antibodies has developed to the point where it is now possible to generate fully human antibodies.

The next challenge was developing an economic manufacturing system. This was not easy because most hybridoma cell lines have low productivity and Mabs need to be given in high doses for successful therapeutic outcomes, needing to be administered in grams as opposed to the milligram quantities of other drugs. The annual production requirement for Mabs can run into hundreds of kilograms or even tons. In 2014 it was estimated that the total demand for antibodies would reach 13 metric tonnes per annum in 2016, a doubling since 2010.[19]

Initially the concentrations of Mabs achieved in reactors were very low and it was obvious that huge improvements in productivity were required to address the demand. The complexity of mammalian cell culture ensured that there would not be an instant answer to this problem, but steady progress was made and, over two decades, antibody titres, or concentrations, increased from tens of milligrams to several grams per litre.

Much of the progress resulted from improvements in the culture media and feeding strategies used to grow cells. Media for animal cell culture are much more complex than those used to grow microbes and contain forty or more individual components including amino acids, vitamins, glucose and salts. In the early days it was also necessary to add animal serum to compensate for nutrient deficiencies that, at that stage, had not been fully resolved. Gradually, as knowledge increased it was possible to eliminate animal-sourced material and use completely chemically defined media components. This not only simplified process development but also removed the danger of animal materials potentially introducing undesirable infectious agents into the process.

In addition to the basic culture medium, small volumes of concentrated nutrients are commonly fed during the culture process; so called 'fed batch culture', a system also widely used in microbial culture. Optimisation of media and feeds has significantly increased the number of cells in a reactor and the duration of the culture, resulting in higher titres. Improvements have also been made in gene expression technology and the screening process for highly productive cell lines has become more effective.

2.4 MANUFACTURING CHALLENGES: PROCESS TECHNOLOGY

It will be no surprise, given the complexity of proteins, that the development of a manufacturing process is a challenging and lengthy process. Depending on the product it can take at least a year and often more, from starting to create a production cell line to being in a position to make the first material for clinical trials. A typical production process flow is shown in Figure 2.3. It should be born in mind that this is a highly simplified view.

In very simple terms the production of a therapeutic protein can be split into an upstream process, which involves the growth of the protein-producing cells to the scale required, and a downstream process, where the protein is recovered and purified to the level required for human therapeutic use. It must then be formulated and packaged into its final form ready for clinical use. Accompanying all steps of production are a vast array of analytical techniques to ensure that the process is operating within its required boundaries and that the product conforms to its required specifications. Small changes in process conditions can have large impacts on the quality of the product.

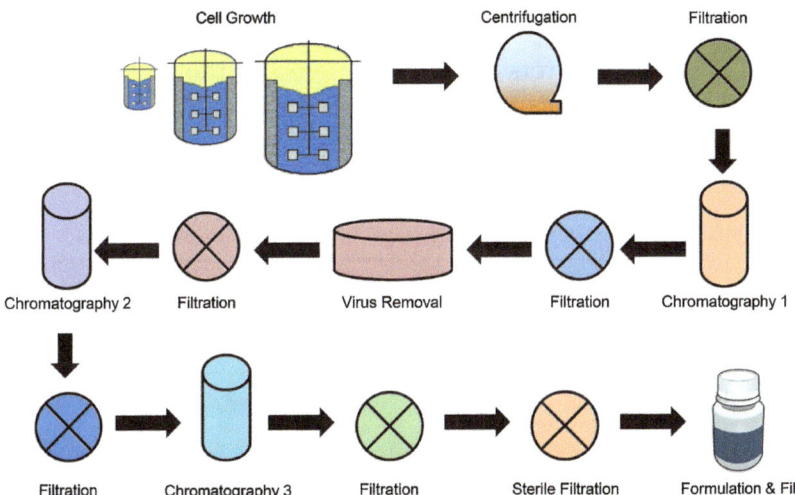

Figure 2.3 Typical biopharmaceutical production process.

2.4.1   Growing Cells to Make Enough Product

It is one thing to demonstrate that a particular protein can be made in a flask in the laboratory and quite another to establish a process on a manufacturing scale. The first challenge is to be able to make enough of the product at an acceptable cost. Dependent on the dose size and number of patients, requirements may vary from a few kilograms per annum to several tons for products such as insulin and some Mabs. In the overall context of pharmaceuticals, these are very small quantities. The antibiotic penicillin, also produced by fermentation, for example, is produced in tens of thousands of tons per year, but, of course, this is a much less complex molecule than a protein. A Mab is several hundred times larger than penicillin (Table 2.5) and has the added complication of being glycosylated (having sugars attached).

For those proteins produced on a small scale this can be done in roller bottles,[24] but this is not practical for those products that need to be produced on a much larger scale. For these much larger containers, bioreactors, are used. The bioreactors need to provide the optimal physical and chemical environment for the cells to grow. This is important whether the product is grown in microbial or mammalian cells. To achieve this, it is crucial to control the pH (by addition of an acid or base), temperature and the level of dissolved oxygen in the culture. Oxygen is typically provided by bubbling air into the vessel. Rotating stirrers mix the contents of the vessel. Maintaining sterility is essential in both the design and operation of the bioreactor to prevent contamination from organisms other than the production strain.

Table 2.5   Properties of representative biopharmaceutical proteins.[a]

Product	Molecular weight (Daltons)	Presence of sugars
Insulin	5808	No
Human growth hormone	22 124	No
Erythropoietin	34 000	Yes
Tissue plasminogen activator	59 000	Yes
Antibody (IgG[b])	150 000	Yes

[a]For comparison the molecular weight of penicillin G, a non-protein natural product produced by fermentation is 356.
[b]IgG is the most common class of therapeutic monoclonal antibody.

It was possible to use experience with earlier, non recombinant, products in the design of bioreactors and processes for genetically engineered mammalian cells. Firstly there was knowledge gained in the production of antibiotics and industrial enzymes, which is done in bioreactors with capacities of tens and hundreds of thousands of litres. Secondly, albeit on a more limited scale, useful experience had been gained with mammalian cells making other products. Wellcome, a major British pharmaceutical company, developed a process to manufacture a vaccine against the Foot and Mouth Disease virus using baby hamster kidney cells grown in large stirred bioreactors and as noted earlier, large scale bioreactors had been used to produce Mabs from hybridomas.

The size of bioreactors has developed to the point that mammalian cell cultures can now be routinely grown at scales of up to 25 000 litres and microbial cultures much larger than this. This has been helped by improvements to the growth media as a result of studies of the nutritional requirements of cells. In some cases the medium is continuously fed to the reactor and spent medium and product is continuously removed. This is particularly useful for proteins that are less stable as it reduces the time that the protein spends in the reactor. Fed batch fermentation might last two or three days for a bacterium like *E.coli* and two weeks for a mammalian cell. The difference in culture duration is the consequence of how fast the two organisms grow. *E.coli* can divide every 20 minutes, whereas a mammalian cell may divide every 20 hours.

2.4.2 Making the Right Product

Optimising the conditions in a bioreactor is just one step in the production process. It is also important to produce a product with the desired properties. This may seem obvious, but large complex proteins made in mammalian cells pose subtle challenges. In practice it is usual to see some minor structural heterogeneity in the population of proteins produced in culture. A typical example would be the variation seen in the number or type of sugar molecules attached to the proteins. This is unavoidable with current technology, but it can be tackled by understanding the significance of the variants and then developing a process that will consistently give the preferred mixture. While many variants are irrelevant, others can have a major impact on the efficacy of the product. In some cases the process can be designed to enrich or remove a particular variant. Once the desired composition has been established, attention is paid to ensuring that the product remains the same as the process is scaled up, or transferred to other facilities and is also consistent from batch to batch. This requires a rigorous understanding and control of the culture conditions.

2.4.3   Recovering the Product

Once the product has been grown, steps are taken to recover it from the culture medium. This is governed by where the product ends up in a culture and this varies. In *E.coli* cultures, with a few exceptions, the product accumulates in the cells, whereas in mammalian cell culture the product is secreted into the surrounding culture medium. Within the context of bacteria, cells are typically separated from the culture broth by centrifugation and then disrupted to release the product, which may be in a soluble state or in some instances present as insoluble granules called inclusion bodies. The granules are made up of aggregated protein in a denatured 'scrambled' state. In this situation recovery involves the use of chemical reagents to make the granules soluble, followed by treatments which allow the protein to refold into its natural state. Life is a little simpler with mammalian cells; cells are removed by centrifugation or filtration and the product is purified from the clarified culture fluid.

2.4.4   Product Purification

Once the product has been recovered the next challenge is to purify the protein. The main impurities that need to be removed arise from the cells used for production. This will be an assortment of cellular debris, host-cell proteins and other substances, such as DNA and chemicals from the growth culture medium. It is important to separate the product from these other substances because they may have undesirable effects. Non-product proteins, for example, may activate an unwanted immune response in the patient. In addition to foreign proteins it may also be necessary to remove unwanted versions of the product; perhaps degraded material or aggregated complexes of the protein, which again could provoke undesirable immune responses.

Most approaches to purification depend on chromatography, a technique in which material is passed through columns containing solid porous beads. A range of beads are used, being chosen to exploit the different physical characteristics of the product to achieve its separation. This is governed by the ionic charge on the molecule, its size, or surface hydrophobicity (tendency to repel water). Product or contaminants may be preferentially bound or retarded as they are passed through these columns, depending on the nature of the product and contaminants. In some cases it is possible to use separation materials which have a very specific affinity for the product molecule and bind it as it passes through the column. A good example of this is Protein A, a natural product produced by a bacterium, which binds specifically to antibody proteins. Antibodies can be purified by passing them through a column of Protein A immobilised on a solid support. The antibodies attach to the material in the column, while contaminants pass through. A solution can then be passed through the column which removes the antibody. Usually, even when an affinity binding method is available, it is necessary to have multiple purification steps using different techniques to achieve the stringent levels of purity required.

When products are made in mammalian cells there is an additional requirement. Animal cell products have been used safely for many years, but safety cannot be taken for granted. Cells have the potential to harbour potentially hazardous viruses and extensive testing is carried out to ensure that such viruses are absent. As an additional safeguard steps are included in the purification process that would remove viruses if present. The chromatography steps provide a level of clearance and it is also usual to include specific virus removal steps, for example using special filters.

2.4.5 Enhancing the Product's Medicinal Effect

There is increasing interest in using chemical steps to improve the medicinal properties of the product. One of the earliest modifications made was 'pegylation', the name given to the process which involves chemically coupling a protein with a synthetic polymer polyethylene glycol (PEG). Developed in the 1970s, this process helps reduce the speed with which some protein therapeutics clear from the body which makes the drug longer acting.[25] Examples of pegylated products include alpha and beta interferons and an antibody fragment (Cimzia®).

Another chemical step, which is generating a lot of excitement in the field of cancer therapy, is to couple a cytotoxic anticancer chemical to an antibody protein which is designed to target cancer cells and leave healthy cells untouched. The aim is to produce an antibody treatment which is more potent than the antibody alone whilst reducing the side effects normally associated with the use of cytotoxic agents in chemotherapy. Two products are marketed (Adcetris® and Kadcyla®) and more are in development.

2.4.6 Preparing the Final Product

Following the purification of a protein, the product is prepared for use in the clinic. This involves a detailed investigation of the best storage conditions for the product. The drug needs to be formulated to ensure its stability for the duration of its desired shelf life. This may need the addition of substances to the solution of product to enhance its stability, such as albumin and various amino acids and carbohydrates. After this the product is put through sterilising filters and aseptically transferred into sterile final product containers at the required dose. The final product is either held as a solution, or in some cases freeze-dried. Once made the product is subjected to a wide range of testing to check that its molecule has the desired characteristics specified.

2.5 AN EVOLVING INDUSTRY

As we have seen the modern biopharmaceutical industry is underpinned by a number of key technologies. These technologies have driven the types of product available and the ways in which they are manufactured and tested. However, the biopharmaceutical industry continues to change and manufacturing technologies evolve to match this.

2.5.1 Economic Factors—Cost of Goods

One of the key drivers of change in the biopharmaceutical sector is the cost of production. The pressure to reduce the manufacturing cost is increasing. One of the key drivers is the patent expiry of the first generation recombinant proteins and the large number of companies looking to produce 'biosimilar' (generic) versions of these products. Even the new generation of innovative products are likely to face greater reimbursement barriers than have been experienced in the past because of the increasing demands on healthcare systems, more therapy choices and the need to demonstrate clear benefit for cost. All these factors will ensure companies look to reduce manufacturing costs.

2.5.2 Geographic Factors—A Global Industry

While the industry is coming under severe price pressure in its traditional market, which up to now has been in the developed world, it is also facing demands to make their products more accessible to patients in the emerging markets of places like China and India. Adapting to the requirements of emerging markets poses particular challenges. Here the price pressure on biopharmaceuticals is particularly acute, one of the influencing factors being the fact that the regulatory systems in such places are much more open to the approval of biosimilar products. Added to this, western biopharmaceutical companies are coming under pressure to develop manufacturing facilities in these new territories so as to sell there. All these factors increase the need for cheaper production facilities.

2.5.3 Scientific Factors—New Products and Stratified Medicine

Traditionally the biopharmaceutical industry has produced a limited number of molecule types, but an ever increasing diversity of product types are now being designed for greater potency and patient safety. In addition to chemical modifications of proteins there are also efforts to re-design existing proteins (*e.g.* alternative antibody formats which are smaller and simpler) and to develop entirely new structures (*e.g.* new binding scaffolds with antibody-like properties).

The increased product diversity is forcing manufacturers to develop new types of unit operation and technology. Critically these need to be capable of handling high-potency products. Manufacturing plants are also having to become more flexible such that they can be rapidly reconfigured to produce different types of product.

New pressures are also being felt as a result of the emergence of stratified and personalised therapies.[26] Personalised medicine represents the ultimate form of stratification, whereby treatment is customised according to an individual patient's genetic profile. Stratified medicine uses the genetic make-up of patients to identify particular groups that will respond to a specific treatment regime. Where stratified medicine is likely to have an impact is in the need for the development and manufacture of a greater number of products and product formulations, many of which will be specific for a smaller patient group and so will only be required in limited quantities. This is likely to bring about broad changes in drug discovery through to patient supply.

2.5.4 Choice of Organisms for Future Production

For the next decade therapeutic proteins are likely to be made in the systems that are standard today, but this may change as a result of product evolution. For example, today there is a great deal of interest in developing smaller, simpler proteins to substitute for some of the very large molecules, especially antibodies, currently in use. There are already one or two products based on antibody 'fragments' and more of these engineered molecules based on antibodies and other proteins are in development.[27]

A major advantage with the simpler molecules is that they can be produced in microorganisms. This means that production in mammalian cell culture may become less dominant. One of the attractions of microbial culture systems is that they are inherently simpler and faster to develop. Advances in genetic engineering may also expand the capabilities of microorganisms. As we saw earlier, one reason to use mammalian cells is because they can put the correct sugars on proteins. In recent years it has been shown that yeasts can be genetically altered to mimic this aspect of mammalian cell biology and we can expect further advances that will make it possible to make more complex mammalian proteins in microorganisms.[28]

2.5.5 Future Manufacturing Plants

One of the major contributors to the price of biopharmaceuticals is the capital cost of building manufacturing plants. A traditional biopharmaceutical plant costs from tens of millions US$ for mid-sized facilities to hundreds of millions US$ for large facilities.[29] Part of this reflects the fact that it typically takes between 3 and 5 years to bring a manufacturing plant into operation. Such construction needs to begin while the product is still in development so that it can be ready for initial manufacture of a product following approval. This means that large amounts of capital must be invested in plant construction before the product has been proven successful.

Traditionally biopharmaceuticals have been manufactured in large, central facilities based on fixed, reusable equipment, much of it constructed from expensive materials such as high-grade stainless steel. In order for this equipment to be reusable it is necessary for the manufacturing facility to be equipped with systems to clean and sterilise in place with steam. These needs drive the high capital cost associated with traditional facilities and require high energy and water use.

Over the last 10 years manufacturing facilities based on mammalian cell culture have increasingly moved toward the use of single-use, disposable equipment based on plastic vessels. Initially single-use equipment was mainly for purposes such as product hold and preparation of culture media and purification reagents but has evolved such that there is now the possibility for fully disposable-based processes including all upstream and downstream process steps. A big step forward in the development of single-use manufacturing technology has been the availability of single-use bioreactors based on the stirred tank design and at sizes suitable for production purposes. These bioreactors are available from a number of different suppliers and have rapidly been adopted by the industry, particularly for mammalian-cell-based production. One factor influencing acceptance has been the improvements seen in culture productivity. For many products, a 1000 or 2000 litre reactor is now a

feasible scale for manufacturing, whereas much larger steel reactors would have been required in the past.

The new technology is promoting a change in the way biopharmaceutical production facilities are designed and operated,
from being traditionally large-scale, fixed and expensive facilities
to more flexible and cheaper plants.[30] The value of single-use
reactors can be seen from the case of Amgen, an American
company, which recently opened a new facility in Singapore
using single-use bioreactors. This was built in less than two
years; half the time required for a conventional biomanufacturing plant.[31] The new plants also have a greatly improved en-
vironmental footprint. This results from the significant
decreases in use of water, land and energy.[32]

The single-use principle is also being applied to other areas of
the manufacturing process and may become more prevalent in
microbial manufacture. In addition to disposable technology,
companies continue to investigate other ways to make biopharmaceuticals more cost effective, for example, by replacing
the current batch processing steps with continuous processes.[33]

Whilst the scale of operation of single-use equipment does not
match that of traditional plants, it is better suited to the production of smaller batch sizes of stratified therapies and production from modern highly productive cell lines; where scale of
production is required this can be achieved through the operation of multiple production trains in parallel.

2.5.6 Process Analytical and Control Technology

What has also proved crucial to the improvement of production
technology has been the development of better analytical techniques to monitor and analyse the process in real time and use
this information to control processing conditions. A major part
of this has been the availability of better sensor technology
suitable for the monitoring of specific metabolites (*e.g.* glucose
level in the reactor), dissolved gases (*e.g.* CO_2, O_2) or physical
conditions (*e.g.* pH, temperature). Much of this sensor technology is small and cheap enough to be considered disposable and
in many cases can even be incorporated into the manufacture of
disposable processing materials, for example single-use bioreactors with sensors incorporated during manufacture which

communicate remotely with external monitoring equipment. A further advantage of this approach is the need for a reduced number of intrusions into the process, thus reducing the chance of contamination.

2.6 CONCLUSION

Biopharmaceutical manufacturing has developed from small beginnings to its current status as a multibillion dollar business behind some of the most advanced therapeutics available. It is an integral part of the business plans of the world's largest pharmaceutical companies and looks to be essential to meeting the healthcare needs of an expanding and ageing global population.

The success of biopharmaceuticals is obvious for all to see. In 2013 the Pharmaceutical Research and Manufacturers of America association reported over 900 proteins in clinical development in the USA alone.[34] Perhaps less obvious have been the remarkable developments in technology which make it possible to manufacture these products on an industrial scale. It has been a truly multidisciplinary effort involving a broad range of scientists and engineers who have created the process, from production organism to manufacturing plant. The story is far from over; with increasing demand for current products and many novel therapies on the horizon. What is clear is that there will be a continuing need for innovation in process development and manufacturing.

REFERENCES

1. PRMLive Top Pharma List, http://www.pmlive.com/top_pharma_list/Top_50_pharmaceutical_products_by_global_sales.
2. J. F. Morrow, S. N. Cohen, A. C. Y. Chang, H. W. Boyer, H. M. Goodman and R. B. Helling, *Proc. Natl. Acad. Sci. U. S. A.*, 1974, **71**, 1743.
3. D. V. Goeddel, D. G. Kleid, F. Bolivar, H. L. Heyneker, D. G. Yansura, R. Crea, T. Hirose, A. Kraszewski, K. Itakura and A. D. Riggs, *Proc. Natl. Acad. Sci. U. S. A.*, 1979, **76**, 106.
4. I. S. Irving, *Science*, 1983, **219**, 632.

5. E. Gebel, Making Insulin, a behind-the-scenes look at producing a lifesaving medication, Diabetes Forecast, Jul 2013, www.diabetesforecast.org. Accessed Jun 2016.
6. A. Philippidis, *Genet. Eng. Biotechnol. News*, 2016.
7. F. Klein, R. T. Ricketts, W. I. Jones, I. A. DeArmon, M. J. Temple, K. C. Zoon and P. J. Bridgen, *Antimicrob. Agents Chemother.*, 1979, **15**, 420.
8. P. Srivastava, P. Bhattacharaya, G. Pandey and K. J. Mukherjee, *Protein Expression Purif.*, 2005, **41**, 313.
9. R. M. Alldread, J. R. Birch, H. K. Metcalfe, S. Farid, A. J. Racher, R. J. Young and M. Khan, Large Scale Suspension Culture of Mammalian Cells, in *Industrial Scale Suspension Culture of Living Cells*, John Wiley & Sons, 2014, ch. 3.
10. R. Harrison, *Anat. Rec.*, 1907, **1**, 116.
11. K. K. Sanford, W. R. Earle and G. D. Likely, *J. Natl. Cancer Inst.*, 1948, **9**, 229.
12. J. H. Tjio and T. T. Puck, *J. Exp. Med.*, 1958, **108**, 259.
13. T. Puck in *Molecular Cell Genetics*, ed. M. M. Gottesman, John Wiley & Sons, 1985, ch. 2, pp. 37–264.
14. K. P. Jayapal, K. F. Wlaschin and W.-S. Hu, *Chem. Eng. Prog.*, 2007, **103**, 40.
15. G. Walsh, *Nat. Biotechnol.*, 2014, **32**, 992.
16. C. Morrison, *Nat. Biotechnol.*, 2016, **34**, 129.
17. N. Yamane-Ohnuki, S. Kinoshita, M. Inoue-Urakubo, M. Kusunoki, S. Iida, R. Nakano, M. Wakitani, R. Niwa, M. Sakurada, K. Uchida, K. Shitara and M. Satoh, *Biotechnol. Bioeng.*, 2004, **87**, 614.
18. R. G. Ham, *Proc. Natl. Acad. Sci. U. S. A.*, 1965, **53**, 288.
19. M. M. Cox, *Vaccine*, 2012, **30**, 1759.
20. C. Sheridan, *Nat. Biotechnol.*, **34**, 117.
21. G. Köhler and C. Milstein, *Nature*, 1975, **356**, 495.
22. J. R. Birch, R. Boraston and L. Wood, *Trends Biotechnol.* 1985, 3(5), 162–166.
23. L. Marks, *The Lock and Key of Medicine: Monoclonal Antibodies and the Transformation of Healthcare*, 2015, Yale University Press, pp. 83–165.
24. E. I. Tsao, M. A. Bohn, D. R. Omstead and M. J. Munster, *Ann. N. Y. Acad. Sci.*, 1992, **665**, 127.
25. P. Wonganan and M. A. Croyle, *Viruses*, 2010, 2(2), 468–502.

26. N. A. Meadows, A. Morrison and D. A. Brindley, *Pharmacogenomics J.*, 2015, **15**(1), 6.
27. C. Enever, T. Batuwangala, C. Plummer and A. Sepp, *Curr. Opin. Biotechnol.*, 2009, **20**, 405.
28. T. I. Potgieter, M. Cukan, J. E. Drummond, N. R. Houston-Cummings, Y. Jiang, F. Li, H. Lynaugh, M. Mallem, T. W. McKelvey, T. Mitchell, A. Nylen, A. Rittenhour, T. A. Stadheim, D. Zha and M. d'Anjou, *J. Biotechnol.*, 2009, **139**, 318.
29. S. S. Farid, *J. Chromatogr. B: Anal. Technol. Biomed. Life Sci.*, 2007, **848**, 8.
30. R. Eibl, S. Kaiser and R. Lombriser, *Appl. Microbiol. Biotechnol.*, 2010, **86**(1), 41.
31. Amgen opens next-generation biomanufacturing facility in Singapore, Amgen news release, 2014 Nov 20.
32. A. Sinclair, L. Leveen, M. Monge, J. Lim and S. Cox, *Biopharm. Intl.*, 2008 (Supplement), 4–15.
33. M. S. Croughan, K. B. Konstantinov and C. Cooney, *Biotechnol. Bioeng.*, 2015, **112**, 648.
34. PhRMA report 2013, Medicines in development, *Biologics*.

Vaccines: The Recombinant Revolution

BARRY C. BUCKLAND

University College London, UK
Email: bucklandbarry@gmail.com

3.1 LEGACY VACCINES

When we receive a vaccine, even one developed many years ago such as DTP, we acquire a gift of good health for a very modest price. In this particular example we gain freedom from the diseases of diphtheria, tetanus and pertussis.

The company making that vaccine and government regulators are well aware of their responsibility to make sure that each and every dose is uniform and will be efficacious. At the same time as we watch the syringe plunger going down we want to know the vaccine is safe.[1] The safety requirements of any vaccine is unusually high, driven by the fact that it will be administered to healthy individuals, including children, unlike a typical drug which is designed to treat patients suffering from a disease.

In the past vaccines were developed by modifying an infectious virus or microbe in some way to make it no longer dangerous and then injecting it to provide immunity from that particular

Engineering Health: How Biotechnology Changed Medicine
Edited by Lara V. Marks
© The Royal Society of Chemistry 2018
Published by the Royal Society of Chemistry, www.rsc.org

infection. This was achieved by weakening a virus so that it is no longer virulent and/or by inactivating the virus or cell. In the early days there was little emphasis on purity. In this way very successful vaccines such as MMR (measles, mumps and rubella), polio, pertussis, hepatitis A, rabies, chicken pox and shingles were developed. Tetanus and diphtheria were different because scientists realized that harm from the infection was the result of a toxin made by the bacteria so the vaccine was based on a modified form of the toxin.

The problem with this empirical approach is that the development cycle can be long, often with a few false starts resulting from safety issues. Even when successful, there is the challenge of reliably manufacturing a product that is poorly defined and potentially variable, which can affect its efficacy and safety. Regulatory agencies and manufacturers have mitigated this risk by demanding that every small detail in the process be rigorously defined and any deviation, even a small one, has to be justified and tracked through a change control process.[1] This conservative approach has resulted in a situation whereby processes developed 40 years ago (such as for the measles vaccine)[2] continue to be carried out in more or less the same way today.

Vaccine production, however, stands on the cusp of change, fuelled by advances in biotechnology.[3] This is not only helping to identify specific disease targets, but also to tailor better vaccines based on an improved scientific understanding of that disease and the required immune response to prevent its infection. Indeed, the emergence of biotechnology has revolutionised what is possible in vaccines, opening up a new path to more effective and safer new products, thereby providing the means to greater freedom from disease.

3.2 HEPATITIS B VACCINE

The first successful application of biotechnology to vaccines was to counter hepatitis B, a disease caused by a virus that can cause debilitating inflammation to the liver and in some cases liver cancer. The virus is one out of five in the hepatitis virus family. It is highly contagious and spread by contact with infected blood, semen and other bodily fluids. Viral hepatitis is one of the most common infectious diseases. While the majority of adults who come into contact with hepatitis B are able to clear it and will recover from an infection without knowing they had it, approximately five percent develop chronic hepatitis B. Many of those with the chronic condition may have no noticeable symptoms for many years but go on to suffer serious liver damage, resulting in their premature death. The younger a person is that picks up the virus, the higher their chance of spreading it on to others. In 2015 WHO estimated that there were approximately two billion people who had been infected with the virus worldwide and 240 million were chronically infected.[4]

In terms of recognition, hepatitis B is a fairly recent disease, first having been identified as a separate entity in the 1950s and shown to be caused by a contagious agent in the 1960s. For many years hepatitis B attracted little in the way of resources in the developed world, being seen as primarily a public health problem of low-income countries or those regarded as social outcasts. Concern, however, began to mount from the late 1960s after a number of renal dialysis units experienced a series of serious hepatitis B outbreaks. These not only struck patients, but also healthcare workers.[5–7]

The first effective hepatitis B vaccine, Hepatavax-B®, was approved in 1981. It was an extraordinary achievement for the time, heralding a new approach to vaccine production, which was no longer reliant on the use of whole viruses or bacteria that had either been killed or weakened. What was new about the Hepatavax-B® vaccine was that it used sub-units or particles of the virus isolated from blood.[7,8]

The development of Hepatavax-B® rested on several earlier breakthroughs. One of the most important was that of Baruch Blumberg,[9] a geneticist based at the US National Institutes of Health who, in the 1960s, discovered a protein in some blood samples taken from an Australian aborigine that turned out to be a surface component of the virus. Initially the protein Blumberg found was called the 'Australian antigen', but the label was later changed to hepatitis-associated B surface antigen (HBsAg). In order for the hepatitis B virus to infect the liver it must first bind to liver cell proteins *via* the HBsAg antigen that sits on the surface of the virus. People make antibodies against this viral surface protein to prevent the virus from attaching. But the virus is tricky and in fact makes far more surface protein than it needs which serves as a decoy for the immune system.[2]

Early on, Blumberg's protein was seized upon as a marker for detecting the virus and by the early 1970s had been adopted as a tool for screening blood for transfusion.[8] At the same time scientists began to explore its use for a vaccine. In the 1970s, several experiments demonstrated that HBsAg did not cause hepatitis B but that it could stimulate immunity against it. Soon after this, researchers at the Fox Chase Cancer Center (FCCC) observed that patients who received blood transfusions with antibodies to HBsAg were less like to develop hepatitis B after the transfusions than those whose blood did not contain such antibodies. Based on this Blumberg and Irving Millman, working at the FCCC, devised a protocol for developing a vaccine using HBsAg particles for which the FCCC filed patent in 1969.

Not having the time or resources to pursue the idea, Blumberg and Millman approached Maurice Hilleman, a major pioneer of vaccines at Merck, to develop the vaccine. Merck was one of the few US based companies that remained active in the vaccine field—many companies had moved away from such products because they were seen as unprofitable and because of the increasing risk of litigation with respect to contaminated vaccines, as well the rising development costs due to greater regulatory controls on safety and efficacy.[8–10]

By 1971 Merck had acquired the FCCC patent to start work on the vaccine. It was a risky venture. Firstly, it was unclear whether the US regulatory authority, the Bureau of Biologics, Standards (now the FDA), would approve such a vaccine because it forbade the presence of blood or even traces of blood in manufactured vaccines.[9] Secondly, the starting point for the vaccine was serum from blood taken from people infected with hepatitis B. This was highly dangerous to work with because it contained the live virus, which is highly contagious. Blood from a person infected with hepatitis B contains extraordinarily large quantities of the virus, about 500 million infectious particles per teaspoon.[2] Isolating the hepatitis B surface protein was also not easy because the blood contained large quantities of other proteins. Hilleman and his team developed an extensive purification process which included three inactivation steps; pepsin to degrade proteins, urea to degrade prions, and formaldehyde to destroy the virus (this method had been used successfully for the vaccine against polio). Once made, the vaccine was clinically tested for safety and efficacy in chimpanzees and humans.

Licensed by the FDA in 1981, Hilleman's blood-derived hepatitis B vaccine took time to be adopted in the clinic. As Hilleman recalled, 'When we bought the vaccine onto the market we had one hell of a time: the doctors and nurses did not want to be vaccinated with human blood.'[2] Hilleman knew that his method of inactivation killed all known viruses, but he recognized the difficulty of convincing physicians. Take-up remained low. Its lack of popularity was enhanced by the arrival of acquired immune deficiency syndrome (AIDS) in the mid-1980s and news of its transmission to French haemophiliacs through contaminated blood products. All of this increased fears that the plasma-derived hepatitis B vaccine could lead to cross-infection with other pathogens despite the very effective inactivation steps taken to guarantee its safety.[11] Similar anxieties were expressed about another plasma-based hepatitis B vaccine developed by researchers at the Institute Pasteur in France, introduced onto the market in 1982.

Adding to the woes of the plasma-based vaccines was the fact that production was dependent on a limited supply of blood from hepatitis B carriers. In addition the production process itself was time-consuming and expensive because it required elaborate steps to inactivate the virus and purify the antigen as well as extra analytical and safety measures.[12]

To try and resolve these issues by the late 1970s scientists had begun to explore the possibility of generating a hepatitis B vaccine using genetic engineering. Also known as recombinant DNA technology, this was a technique developed through the combined efforts of Paul Berg, Janet Mertz, Herbert Boyer and Stanley Cohen in the early 1970s. The method enabled the transfer of genes from one organism to another, providing the means to transform bacteria into 'factories' to produce foreign proteins (see Figure 2.1 in Chapter 2 by Alldread and Birch). This approach was highly attractive for producing a hepatitis B vaccine. Importantly, it offered a way to obtain the hepatitis B surface protein without the use of blood and to produce large quantities of the vaccine.[8,9,12]

The first recombinant hepatitis vaccine to hit the market was Recombivax HB®, approved in 1986, first in Germany and then in the USA. The vaccine had been created through a joint venture between Merck and William Rutter of the University of California, San Francisco, Chiron, a small biotechnology start-up, and Ben Hall from the University of Washington, Seattle. Leadership within Merck was by Roy Vagelos and Ed Scolnick (Figure 3.1).[11]

Work on the project had started in 1977 when Merck sent Rutter material containing the virus. This was used to clone the virus and obtain the genetic sequence of HBsAg, its surface protein. Thereafter the team began inserting the gene from the HBsAg into a variety of host organisms to reproduce the antigen. Initial attempts using *E. coli* were not successful because the bacteria made the wrong form of the hepatitis B surface antigen. Better results were achieved with *Saccharomyces cerevisiae*, otherwise known as baker's yeast. In this case the yeast cells expressed an active surface antigen that evoked the desired immunological response.[11,13]

Getting a good expression of the hepatitis B surface antigen by the yeast cell had been a major challenge and was aided by

Figure 3.1 Ed Scolnick, previously President of the Merck Research Laboratories, provided the vision and support for the development of remarkably successful vaccines against Hepatitis B and HPV.

breaking open the yeast cell without damaging the surface protein using methods pioneered at University College London[14]
and then purifying the hepatitis B surface antigen away from
other yeast proteins, cellular debris and DNA. Critically, a
chemical modification to the antigen was required to make the
particle immunogenic and this was achieved by treatment with
thiocyanate.[15]

In parallel to the development of Recombivax HB®, a similar
vaccine was developed by Smith Kline Beecham in Belgium. This
was done in collaboration with the researchers Ken Murray,
based at Edinburgh University, and Peter Hans Hofschneider,
based at the Max Planck Institute in Martinsried, and the
support of Biogen, a newly formed biotechnology company.[7] It
was approved for market by the FDA in 1989 with the trade name
Engerix-B®.

The recombinant vaccine represented a major advance.
Importantly, it could induce an immune response without the
risk of infecting recipients with the hepatitis B virus. Moreover, it
had a shorter production cycle (12 instead of 65 weeks). In
addition, it was consistent between batches and its supply was
continuous. Yet, it also had a downside; its price was higher than
those of the plasma vaccines. This was in part because it was
patent-protected and only a limited number of companies produced it. Despite the high price, the recombinant vaccines soon
pushed the plasma vaccines off the market in North America and
Western Europe.[16]

Initially the recombinant vaccines were directed towards
health care workers. This however shifted over time once its
remarkable safety had been demonstrated. Soon the target
population included every child in the world. The hepatitis B
vaccines are still in use today and have by now protected many
hundreds of millions of people. To give a sense of the magnitude
of the disease, in the 1980s, around 120 million carriers of
hepatitis B were reported every year in China and 10% of
newborns acquired chronic hepatitis B from their mothers. The
disease was responsible for almost half a million deaths from liver
cancer and end-stage cirrhosis each year before China's universal
hepatitis vaccination programme was implemented in 1992.
Official research in 2006 showed that infant hepatitis B carriers
have been reduced from 9.7% in 1992 to less than 1% in 2006 in

China. Also, the carriers in the total population dropped from 10% to 7.2%.[9]

The Chinese programme was aided by the fact that in the early 1990s Merck struck a deal with the Chinese government to transfer its expertise and the technology to manufacture the recombinant vaccine to a plant in Shenzhen. The aim was to provide enough hepatitis B vaccine to immunize all the babies in China. This was initiated by Vagelos, then chairman of Merck, Sharpe & Dohme (known as Merck in the USA and Canada).[2] The motivation for this arrangement was primarily driven by public health considerations. There was negligible financial return for Merck. Nonetheless, it helped bolster the company's reputation at a time when the pharmaceutical industry was generally coming under fire for the high prices of drugs.[17]

By the mid-1990s patents on the first recombinant hepatitis B vaccines began to expire. This opened the way to manufacture of generic versions of the vaccine and many vaccine companies based in India have been developing their own versions of the hepatitis B vaccine since 2000, some of which have been approved by the Indian government. Some of these vaccines, produced by companies such as Serum Institute and Biological E, have been certified by the World Health Organization (WHO), which has reassured many other countries regarding their quality. These vaccines are sold at an incredibly low cost (below 40 cents per dose) internationally and used by all age groups. As an example, Biological E, based in Hyderabad and led by Mahima Datla, sells more than a hundred million doses of hepatitis B vaccine in many different countries each year either as a stand-alone vaccine or in combination with other vaccines.

3.3 HPV VACCINE

Soon after completing its development of the recombinant hepatitis B vaccine Merck turned its attention to a recombinant vaccine against the human papillomavirus (HPV). There are more than 100 types of this virus. Most of them cause no symptoms and disappear on their own. Others cause harmless growths like verrucas and warts. Yet, the virus can also cause cervical cancer, a disease that globally kills 260 000 women every year, 85% of them being in developing countries, where it is a leading cause of death among women.[18]

Merck's decision to develop a vaccine against HPV was a big gamble. When the company first began work on the vaccine, in early 1992, the link between HPV and cervical cancer remained heavily contested, although scientific and epidemiological evidence accumulated from the early 1970s strongly implicated HPV as the etiological agent in causing cervical and other cancers. One of the key figures who championed the link was Harald zur Hausen, a German virologist, who came to this conclusion after conducting extensive DNA analysis of biopsies taken from cervical cancer patients. This revealed two types of HPV responsible for the cancer, HPV 16 and 18. zur Hausen would later be awarded a Nobel Prize, in 2008, for his work. Yet, the medical world took time to accept his findings because for a long-time it had been assumed the cancer was caused by the herpes virus. The resistance was so great that zur Hausen struggled to find a pharmaceutical company willing to develop a vaccine against HPV. An additional problem was such a vaccine was seen as unprofitable.[19]

Attempts to make vaccines against papilloma viruses went back many decades, inspired by the results of experiments in the 1930s, which demonstrated that domestic rabbits developed immunity after being inoculated with the cottontail rabbit papilloma virus. Progress, however, was limited, hampered by the fact that it was difficult to grow the viruses on cultured cells. Instead researchers had to rely on infected animal tissue, which was infectious. The situation changed with the arrival of recombinant DNA, which paved the way to cloning the viruses in bacteria.[20]

The development of the recombinant HPV vaccine rested on the work of several groups of scientists who, in the 1990s, discovered that when L1 and L2, the two proteins that make up the HPV, were genetically manipulated into the form of a soccer ball they could generate an HPV particle that resembled the real virus and was capable of evoking an immune response to the targeted HPV types. This opened the possibility of developing a vaccine without the need to use the viral infectious DNA. Two of the groups involved in this work were Ian Frazer and Jian Zhou at the University of Queensland in Australia and John Schiller and Douglas Lowy at the National Institutes of Health, USA.[21]

In 1992 Merck licensed the method developed by Zhou which allowed them to generate the recombinant HPV virus-like particles L1 and L2 using the vaccina virus and animal cells. This was to be the start of a challenging project.[1,22,23]

Fortunately, both Edward Scolnick and Kathrin Jansen within the Merck Research Laboratories were both convinced that developing an effective vaccine based on virus-like particles (VLPs) would be possible. Scolnick and Jansen not only provided extra-ordinary skilled leadership to the programme, but also made sure the necessary resources were available over a long time period to support a project of such magnitude.

One of the first issues the Merck team faced was the fact that the HPV viruses are fairly large molecules (multimillion Daltons molecular weight) which makes their cloning a special challenge.[24] In fact, the viruses themselves face the same problem, in that if they needed to package the genome coding for a multimillion Dalton entity, they would not have sufficient room inside their protein shell, known as a capsid, for such a large genome. They solved this problem by evolving over thousands of years the ability to produce proteins of modest size (24 kDa and higher), which self-assemble into icosahedral virus particles after viral replication inside a host cell.[25] In appropriate expression systems, these proteins retain that ability to self-assemble into VLP structures whose surface is immunochemically similar to that of the actual virus (Figure 3.2).

By this time the hepatitis B vaccine had proven to be both effective and safe and a similar strategy was pursued for HPV within Merck. A recombinant approach was taken from the start using the same *Saccharomyces cerevisiae* host cell and it was reasoned that the outside protein coat of the virus, L1, could be the foundation of a successful vaccine. The L1 protein self-assembled into VLPs within the yeast and accounted for about 15% of the total soluble protein. A multistep purification process, consisting of microfiltration and cation-exchange chromato-graphy, was developed, yielding highly purified L1 VLP pre-parations that were 98% homogeneous.[24]

Self-assembly of the HPV L1 VLPs takes place within the yeast. Five L1 molecules come together to form a pentamer, often referred to as capsomer. A total of 72 of these capsomers assemble to form both five and six axes of symmetry of the 55 nm diameter

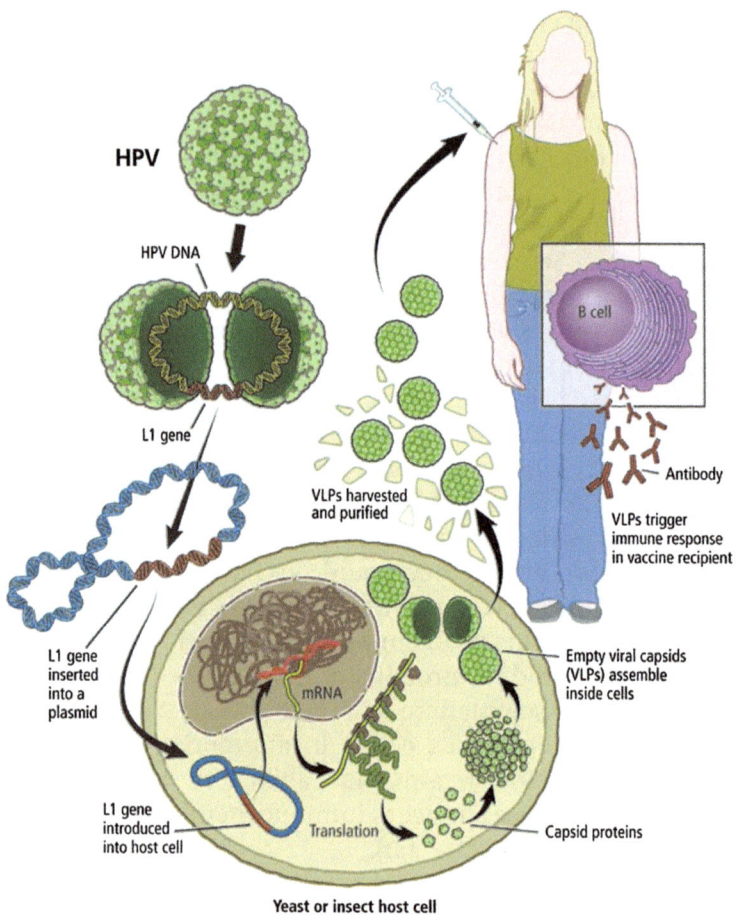

Figure 3.2   Diagram showing how HPV vaccines are produced.
Reproduced from Jin X. W., Lipold L., Sikon A., Rome E. Human
papillomavirus vaccine: safe, effective, underused, *Cleve. Clin. J.
Med.*, 2013, 80, 49–60. With permission from The Cleveland Clinic
Foundation. © 2013 The Cleveland Clinic Foundation. All rights
reserved.

icosohedral HPV VLP structure. Optimally, at the capsomer
interfaces, the L1 molecules are held together by hydrophobic
forces and disulfide bonds. Amino acid carboxylate structures
within the L1 molecules are neutralized when the calcium concentration in the environment is high, enabling highly ordered
tight bonding between L1 molecules within and between

capsomers. Inside the cell, however, calcium concentrations can be low, resulting in reduced disulfide bonds and negatively charged carboxylate structures. Strong charge repulsion can disrupt or prevent uniform VLP assembly.

Utilizing this knowledge and the observation that the intra-yeast-assembled VLPs of HPV 6, 11 and 16 were found to be heterogeneous with a degree of size variation and the presence of irregularly shaped structures, a VLP disassemble/reassemble step controlling the carboxylate switch from charged to neutralized was inserted into the purification manufacturing process for HPV types 6, 11 and 16.[26] The process was optimized to produce reassembled VLPs which had consistent uniformity of size and shape, controlling for higher-order aggregation of VLPs.[27] In this way, more perfect particles were formed.

A key early step in HPV vaccine development is to use these VLP particles to generate antibodies from rabbits. These antibodies then become the basis of an *in vitro* potency assay, or test, which, if developed correctly,[26] is used as a predictor of clinical performance. This provides a foundation for a series of batch release tests to re-assure the manufacturer and the government regulatory agency that each lot will be efficacious.

One consequence of the intracellular production of the VLPs was the necessity to disrupt the yeast cells to perform lab-scale purification to determine both the quality and quantity of the VLPs. As this processing became a limitation for process development, Marc Wenger and colleagues, in collaboration with the Merck Research Laboratories Bioprocess team and the Biochemical Engineering Group at University College London developed a microscale version of the purification process. As an example, a process was developed such that the chromatography could be completed at the microliter scale in a micropipette column. Using this micro method in process development, along with analytical development, allowed high screening throughput to be achieved and the rate of process development significantly improved.[28]

For success, progress in process development needs to be integrated with the clinical development programme and technology transfer to manufacturing.[1] As changes are made we need to predict how these will impact clinical performance. The clinical programme dominates and is the final confirmation that everything is in place for the manufacturing process. Imagine the suspense! In the case of Gardasil® those of use who were involved in the development of the vaccine sat in a lecture room to learn of the results of a 22 000-patient double-blinded clinical study which lasted for over 3 years. Imagine the sense of accomplishment to learn in the first few minutes of the presentation that all the incidences of disease were in the placebo group and none in the vaccine group. The product, Gardasil®, made by Merck (a four-valent HPV vaccine) was licensed by the FDA in the USA in 2005 for the prevention of cervical cancer. A year later another vaccine, Cervarix®, made by GSK (a two-valent HPV vaccine) received European approval.[29,30]

3.4   INFLUENZA VACCINE (FLUBLOK)

Influenza (or flu) is a highly contagious, acute respiratory disease, caused by influenza viruses, that occurs seasonally in most parts of the world. Epidemics occur annually and are the cause of significant morbidity and mortality worldwide.[31] Influenza affects all age groups and in the USA alone, 25 to 50 million people contract influenza each year. The symptoms of the flu are similar to those of the common cold but tend to be more severe. Fever, headache, fatigue, muscle weakness and pain, sore throat, dry cough and a runny or stuffy nose are common and may develop rapidly.

Biotechnology is now poised to help improve the development of influenza vaccines. The 'flu vaccine' presents two key challenges. The first is that it is a seasonal vaccine; it changes every year. The second is that new influenza A viruses can arise as a result of a re-assortment of viral RNA molecules between virus strains ('antigenic shift'), resulting in a new subtype of the virus. This re-assortment introduces a new protein, hemagglutinin (HA), into the circulating viral strains, which can result in a pandemic with the potential for an international public health catastrophe such as the 1918 Spanish Flu. Either way, response needs to be rapid. In the first instance a manufacturer must get ready to have tens of millions of doses available in a timely manner for the next 'flu season'. In the second instance, rapid development of a vaccine can prevent millions of deaths.

Historically, 'flu vaccines' have been based on the actual 'flu virus' which has been adapted to be less virulent. A number of ways of manufacturing the virus-based influenza vaccine have been developed and the most widely used is the classical method of cultivating the virus inside freshly fertilised chicken eggs. This is followed by purification and inactivation.

The 'flu virus' is adapted so that it can propagate to high numbers inside the chicken eggs prior to purification and inactivation. This manufacturing process has served us well for many decades but has several important limitations.

With an increased knowledge of the underlying biology of influenza infection and with the development of recombinant DNA technology it has become possible to create an influenza vaccine without handling the actual virus.[32,33] We now know that the HA protein which coats the influenza virus is the part of the virus that changes every year. We also know that antibodies effective against HA can provide protection from infection.

A team led by Manon Cox at Protein Sciences, a biotechnology company based in Meriden Connecticut, recently developed a 'flu vaccine' based on HA protein which was demonstrated to be at least as efficacious as the virus-based vaccine. In 2013 the FDA approved the vaccine on the basis of convincing efficacy and safety data generated in several clinical studies. Known as Flublok®, this was the first recombinant DNA protein-based influenza vaccine (Figure 3.3).

The breakthrough made with Flubok® greatly facilitates vaccine production, especially in response to a pandemic. For example, when the H7N9 virus was identified two years ago in China it was not long before the sequence was elucidated and published on the Internet. While other groups need the actual virus in order to proceed to develop a vaccine, Protein Sciences can get everything they need from the Internet. It is an amazing reflection of the power of today's technology and represents an extraordinary milestone in vaccine development.

The sequence of the HA protein is published and can be downloaded from the Internet. This can be emailed to a DNA synthesis company, which can create a plasmid of the DNA in a few days. This is then shipped by express mail in a little plastic bag to Protein Sciences who can incorporate this DNA into their standard protein expression platform. Within weeks the manufacturing platform, based on cell culture and baculovirus expression, is cranking out purified HA protein in accordance

Figure 3.3   Manon Cox led the team at Protein Sciences for the successful development of Flublok®, the first and only recombinant-protein-based influenza vaccine approved by the FDA.

with Good Manufacturing Practice guidelines so that it is immediately useful as a vaccine. This platform has the power to revolutionize the making of influenza vaccines in the future, taking vaccine technology into a new era.[34]

3.5   CONCLUSION

Vaccines have provided an unsurpassed health benefit for a very long time. Recombinant DNA technology first made an impact with the hepatitis B vaccine in 1986 and has indeed taken the development of vaccines and the possibilities of vaccination to a whole new level. Besides the examples of hepatitis B, HPV and influenza described here, we have recently seen the development of a whole new wave of vaccines. For example, we have seen the development of a hepatitis E vaccine in China in much the same way as for hepatitis B. Another example is the case of shingles, a protein based vaccine combined with a strong adjuvant developed by GlaxoSmithKline (GSK) which seems to have exceptional efficacy. GSK has also developed a vaccine candidate for malaria based on a protein antigen. Because the new genetically engineered vaccines are, in general, well characterized the opportunities are growing for process innovation. Improvements can be made to the process and bridging based on the concept of analytical comparability.[34]

For legacy vaccines, such as measles, the product itself is made from the virus which has been adapted so that it is less virulent. A vaccine like this is difficult to purify and not easy to characterize analytically. As a consequence the manufacturing process for most legacy vaccines is fixed because of the challenge of proving that a modification in the process has not resulted in a change in the product. Some of these processes are already based on 50 year old technology with no prospect of this changing in the coming years. This is a key fact; we can imagine that technologies to make a recombinant-based vaccine can continue to improve, with a path forward for implementation based on analytical characterization. This will allow for more competition, more modern processes, reduced costs, improved quality and a more reliable supply. We have already seen this happen with the hepatitis B vaccine.

REFERENCES

1. B. C. Buckland, *Nat. Med.*, 2005, **11**(4), S16–S19.
2. P. A. Offit, *Vaccinated, One Man's Quest to Defeat the World's Deadliest Diseases*, Harper Collins, 2007.
3. J. O. Josefsberg and B. Buckland, *Biotechnol. Bioeng.*, 2012, **109**(6), 1443–1460.
4. WHO, Hepatitis B Fact Sheet, No. 204, July 2015, http://www. who.int/mediacentre/factsheets/fs204/en/. Accessed Jun 2016.
5. W. Muraskin, *The War against Hepatitis B*, University of Pennsylvania Press, 1995.
6. J. Stanton, *Hist. Philos. Life Sci.*, 1995, **17**, 113–122.
7. F. Huzair and S. Sturdy, *Stud. Hist. Philos. Biol. Biomed. Sci.*, 2017, **64**, 11–21.
8. M. Patlak, *Beyond Discovery: The Path from Research to Human Benefit. The Hepatitis Story*, National Academy of Science, 2000, http://www.nasonline.org/publications/beyond-discovery/ hepatitis-b-story.pdf.
9. *Hepatitis B: The Virus, the Disease and the Vaccine*, ed. I. Millman, T. Eisenstein and B. S. Blumberg, Springer, 1984.

10. B. S. Blumberg, Vaccine against viral hepatitis and process, US Pat. 2626191 A, filed 8 Oct 1969, granted 18 Jan 1972.
11. R. Vagelos and L. Galambos, *Medicine, Science and Merck*, Cambridge University Press, 2004, p. 167.
12. P. Tiollais and C. Pourcel, *Nature*, 1985, **317**, 489–495.
13. E. M. Scolnick, A. A. McLean, D. J. West, W. J. McAleer, W. J. Miller and E. B. Buynak, *JAMA, J. Am. Med. Assoc.*, 1984, **251**(21), 2812–2815.
14. J. A. Currie, P. Dunnill and M. D. Lilly, *Biotechnol. Bioeng.*, 1972, **14**, 725–736.
15. D. E. Wampler, E. D. Lehman, J. Boger, W. J. McAleer and E. M. Scolnick, *Proc. Natl. Acad. Sci. U. S. A.*, 1985, **82**(20), 6830–6834.
16. L. J. Frost and M. R. Reich, *Access: How Do Good Health Technologies Get to Poor People in Poor Countries?* Harvard University, 2009, p. 73, http://www.accessbook.org/downloads/chapter_4_AccessBook.pdf.
17. Roy Vagelos talks about leadership and the need for new drug pricing policies, Wharton University of Pennsylvania, 2002 June 2, http://knowledge.wharton.upenn.edu/article/roy-vagelos-talks-about-leadership-and-the-need-for-new-drug-pricing-policies/.
18. N. M. Nour, *Rev. Obstet. Gynecol.*, 2009, **2**(4), 24–44.
19. P. McIntyre, *Finding the viral link: the story of Harald zur Hausen*, Cancer World, July–Aug, 2005, pp. 32–37, http://www.cancerworld.org/pdf/6737_cw7_32_37_Masterpiece%20(2).pdf. Accessed Jun 2016.
20. Papillomaviruses and human cancer, 2015, https://rybicki.wordpress.com/2015/03/11/papillomaviruses-and-human-cancer/. Accessed Jun 2016.
21. S. S. Krishnan, *The HPV Vaccine Controversy*, Greenwood, 2008, p. 97.
22. J. Zhou, X.-Y. Sun, D. J. Stenzel and I. H. Frazer, *Virology*, 1991, **185**, 251–257.
23. I. Frazer, A panoramic synthesis, speech for Balzan Prize 2008, http://www.balzan.org/en/prizewinners/ian-h—frazer/rome–21-11-2008-forum-frazer. Accessed Jun 2016.
24. J. T. Bryan, B. Buckland, J. Hammond and K. U. Jansen, *Curr. Opin. Chem. Biol.*, 2016, **32**, 34–47.

25. R. C. Reichman, Human Papillomavirus Infections, in *Harrison's Principles of Internal Medicine*, ed. D. L. Longo, A. S. Fauci, D. L. Kasper, S. L. Hauser, J. L. Jameson and J. Loscalzo, 2012, ch. 178.
26. R. D. Sitrin, Q. Zhao, C. S. Potter, B. Carragher and M. W. Washabaugh, Recombinant virus-like particle protein vaccines, in *Vaccine Analysis: Strategies, Principles, and Control*, ed. B. K. Nunnally, V. E. Turla and R. D. Sitrin, Springer, 2015.
27. H. Mach, D. B. Volkin, R. D. Troutman, B. Wang, Z. Luo, K. U. Jansen and L. Shi, *J. Pharm. Sci.*, 2006, **95**, 2195–2206.
28. M. D. Wenger, P. DePhillips, C. E. Price and D. G. Bracewell, *Biotechnol. Appl. Biochem.*, 2007, **47**, 131–139.
29. I. H. Frazer, G. R. Leggatt and S. R. Mattarollo, *Annu. Rev. Immunol.*, 2011, **29**, 111–138.
30. J. Stephenne, *Vaccine*, 1990, **8**, S69–S73.
31. W. P. Glezen, *Epidemiol. Rev.*, 1982, **4**, 25–44.
32. L. K. Miller, A virus vector for genetic engineering in invertebrates, in *Genetic Engineering in the Plant Sciences*, ed, N. J. Panopaulus, Praeger, NY, 1981, pp. 203–222.
33. M. Cox, *J. Invertebr. Pathol.*, 2011, **107**, 531–541.
34. B. C. Buckland, *Hum. Vaccines Immunother.*, 2015, **11**(6), 1357–1360.

Monoclonal Antibodies: A Revolution in the Transformation of Healthcare†

LARA V. MARKS

University College London, UK
Email: l.marks@ucl.ac.uk

4.1 INTRODUCTION

The rise of biotechnology in medicine is commonly attributed to the scientific breakthrough in genetic engineering in the early 1970s. Yet, it was just as closely tied to another event that occurred in 1975. This was the development of monoclonal antibodies (Mabs). Wherever you look today, Mabs are at the heart of laboratory research to determine the pathways of disease and the diagnosis of patients and at the cutting edge of treatment. Mab drugs are now one of the fastest growing therapeutic areas in medicine, comprising a third of all newly introduced treatments and six out of ten of the best-selling drugs. Their use extends

---

†This chapter draws extensively on ref. 1.

Engineering Health: How Biotechnology Changed Medicine
Edited by Lara V. Marks
Published by the Royal Society of Chemistry, www.rsc.org

across many different diseases, ranging from cancer and auto-immune disorders through to central nervous system problems.[1]

4.2 WHAT IS A MONOCLONAL ANTIBODY

Invisible to the naked eye, Mabs are laboratory-produced molecules derived from the millions of antibodies the immune system continuously makes to recognise and combat foreign invaders such as bacteria, viruses, fungi, pollen or other particles considered alien to the body, such as chemicals and toxins. Millions of different types of antibody can be found in the blood of humans and other mammals. Produced by white blood cells, known as B lymphocytes, the antibodies are large Y-shaped proteins which the immune system uses to identify and neutralise foreign substances that invade the body. Each antibody is highly specific, that is it will only bind to one particular molecule, known as an antigen, found on the surface of a harmful agent. Once antibodies have locked on to their antigen, they and other types of cells produced by the immune system destroy the foreign substance (Figure 4.1).

4.3 UNDERSTANDING THE NATURE OF IMMUNITY

The development of Mabs has a long history and is rooted in the very early medical and scientific efforts to understand the nature of immunity and combat infectious diseases. Right from the

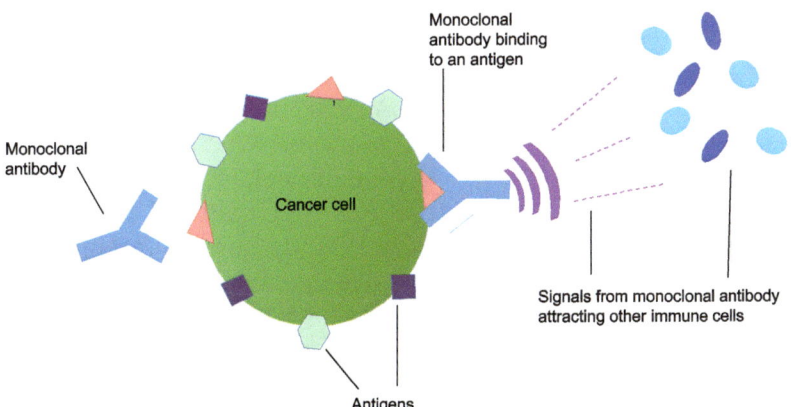

Figure 4.1 Diagram showing monoclonal antibody binding to an antigen.

time of the plague astute observers noted that some individuals who suffered and survived one infectious disease outbreak remained immune the next time it resurfaced. This insight laid the foundation for the development of vaccines, the first of which was directed against smallpox in the late 17th century. Such vaccines involved injecting a weakened form of a disease organism to confer immunity.

Much of the early development of vaccines was done with little understanding of the immune mechanism that underlay their success. Many believed immunity stemmed from the infectious agent itself. By the end of the nineteenth century, however, scientists began to uncover evidence suggesting that immunity occurred due to a process taking place within the body. This understanding entered a new phase as a result of some work conducted by researchers at the Koch Institute in Berlin in 1890. That year Emil Von Behring and Kitasato Shibasaburō discovered that it was possible to cure animals infected with diphtheria and tetanus with blood serum taken from animals that had survived such infections. Moreover, the same serum protected animals with no previous exposure to diphtheria or tetanus. On the back of these findings they proposed that there could be something in the blood of the exposed animals that conferred immunity.[2-4] Soon after this Paul Ehrlich, an expert in structural chemistry, detected a substance in blood that appeared to provide immunity against plant toxins which he called 'antibodies'. These antibodies, he noted, were very specific—antibodies against ricin, a poison found in castor beans, only offered protection against ricin.[5]

By 1897 Ehrlich had a developed a theory which opened up a new understanding of immunity. All cells in the body, he argued, possessed a wide variety of receptors, or what he called 'side chains', which acted like gatekeepers or locks for each cell, only letting in substances with structures that matched their own, such as nutrients for the cell, and keeping out any foreign substances, or antigens, that could destroy the cell. Ehrlich proposed the hypothesis that whenever a cell encountered an antigen it produced more side chains, which broke off to form antibodies able to bind and neutralise any free-floating antigens. Each antibody, he believed, had receptors designed to match specific antigens in the way that a key fits a particular lock. Based on this he suggested that one day it would be possible to develop antibodies that, like a 'magic bullet', would be able to seek out and destroy disease-causing organisms without damaging the rest of the body.[6–9]

The accuracy and details of Ehrlich's theory were debated well into the twentieth century. One of the key puzzles was how the immune system could produce so many different kinds of antibody, each able to specifically target just one of a near-infinite number of antigens that invaded the body. This was not easy to answer because for many years scientists struggled to isolate and purify single antibodies with known targets from the billions made by the body.

4.4 SOURCING ANTIBODIES—THE ROAD TO MONOCLONAL ANTIBODIES

For the greater part of the twentieth century the only source of antibodies were those that could be obtained from blood serum taken from previously immunised animals. This serum, known as antisera, however, had major limitations. Not only did it contain a mixture of antibodies, each of which recognised a different antigen, its supply depended on an individual animal's lifetime. Moreover, batches from one animal could vary. This reflects the fact that animals cannot be stimulated to produce specific antibodies and that their serum contains thousands of different antibodies, each differing in their binding capacity and specificity. All of this made it difficult to get standardised antibodies. Overall the preparation and purification of antibodies was a time-consuming and expensive process.

Over the years scientists devised various techniques to improve the process, including fractionation to separate the antibodies found in the serum and producing artificial antibodies, but with limited success. By the early 1970s several scientists in the USA and Britain had independently managed to produce single antibodies with known specificity, but only in small quantities. Moreover, the antibodies only survived for short periods of time.[10]

The situation all changed in 1975 when César Milstein and Georges Köhler at the UK Medical Research Council's Laboratory of Molecular Biology (LMB) in Cambridge, UK, devised a laboratory technique to produce endless quantities of identical antibodies that could bind to a specific target. This they did as part of their search to find a tool to determine the genetic origins of antibody diversity and specificity. Their method involves several steps. In the first instance, a mouse is immunised against a particular antigen. Following this, the antibody producing cells, lymphocytes, are harvested from the mouse's spleen and fused with a cell associated with myeloma, a type of bone marrow cancer. This creates a hybrid cell that secretes antibodies. The hybrid cell is immortal and can be grown indefinitely, either in the abdominal cavity of mice or in tissue culture, where it will produce unlimited quantities of identical antibodies to a selected target.[11] Such antibodies are called 'monoclonal' because they are made from identical immune cells that are clones of a unique parent cell.

Both Milstein and Köhler were awarded the Nobel Prize for their achievement in 1984. Their technique opened up access to antibodies on an unprecedented scale and a standardised reagent that was more like a chemical than a biological serum product. With Mabs scientists could now experiment with the same reagents and compare results in a way they could never have done before.

4.5   A REVOLUTIONARY TOOL FOR MEDICAL LABORATORY RESEARCH

Originally conceived of by Milstein and Köhler as a laboratory tool, Mabs quickly extended frontiers in research well beyond immunology and revolutionised the way scientists investigated and analysed biological phenomena more generally. Reflecting this, between 1981 and 1983 American National Institute for Health funding for Mabs rose from $78 million for 768 projects to $206 million for 1940 projects. Critically, the technology enabled the detection of unknown molecules and established their function for the first time.[12]

Where Mabs proved particularly useful was for immunohistochemistry. Sometimes known as immunocytochemistry, immunochemistry is a common laboratory technique that exploits the binding mechanism of antibodies to specific components of cells to analyse and identify different cell types. By the

1970s scientists had successfully attached various florescent dyes, radioisotopes and enzymes (especially horseradish peroxidase, a yellowish-brown pigment) to antibodies as a means of staining tissue and identifying various cells in biological samples. A major advantage of the immunochemistry-based tests was that they could be used in conjunction with an electron microscope. Prior to this development these tests were reliant on antisera which were unreliable both in terms of supply and results. The whole situation was transformed by Milstein and Köhler's technique. Importantly it provided access to unlimited standardised antibodies which were highly specific in their target.

By the early 1980s Mabs had opened up a whole new world of research. Not only did they help improve older methods of pathological analysis based on immunochemistry techniques, bringing a new degree of precision and accuracy; they also revealed previously hidden anatomical aspects of the body. Some of their earliest impact was in the exploration of the brain and the central nervous system.[13]

4.6 A NEW TOOL FOR DIAGNOSTICS

In addition to providing a tool for probing previously hidden parts of the body, Mabs were rapidly utilised for immunology-based diagnostic tests. Developed from the late nineteenth century the primary means of detection for such tests was the antigen–antibody reaction. The first such test was for typhoid, introduced in 1896. Many others followed, and by the 1970s such tests were widespread in clinics and hospitals, being used not only for detecting disease and conditions like pregnancy but also to determine blood grouping for blood transfusions. Yet these tests had a major disadvantage—they were reliant on antiserum which greatly reduced their accuracy.

Mabs laid the foundation for the development of simple, cheap, fast, and accurate diagnostics suitable for mass screening and automation. Where they were of particular help was in the detection of infectious disease. One of the key advantages they had was that they enabled the determination of disease directly from a clinical sample, eliminating the need for the onerous and time-consuming culturing of the causative infectious agent. This dramatically reduced the time involved in testing. Some idea of the revolution Mabs brought to diagnostics can be gauged from the tests for human sexually transmitted diseases. Prior to Mabs this had necessitated culturing the suspect microbe, a process that could take up to six days, and conducting a confirmatory test. A Mab test, by contrast, could be done within minutes.[14,15]

By the mid-1980s Mab-based diagnostics had been developed for the detection of various infectious diseases, including hepatitis B, herpes, chickenpox, rabies, legionellosis, influenza and the human immunodeficiency virus. In addition to infectious disease, Mab tests were deployed for detecting a host of other conditions, ranging from cancer to helping to determine a heart attack, typing blood for transfusion and tissue for organ transplants through to uncovering drug abuse within the workplace and in the sports arena. They also became critical components in home-testing diagnostic kits, such as those for pregnancy and ovulation. Biotechnology companies rapidly capitalised on the potential of Mab-based diagnostics. By 1991 the market for such tests was estimated to be worth approximately US$1.9 billion.[15,16]

4.7   THE ROAD TO THERAPEUTICS

While the adoption of Mabs for diagnostics was rapid, their use for therapeutics took much longer. The foundation for their use in therapy had been established as long ago as the 1890s when Behring, with the assistance of Ehrlich, began to apply the new found knowledge about antibodies to find a cure for diphtheria, a disease then claiming the lives of approximately 50 000 German children a year. Their objective was to develop a serum therapy. This involved giving patients blood serum taken from animals immunised against diphtheria. By 1894 they had perfected a standardised, potent serum from horses that proved clinically safe for use in humans. This proved highly successful, dramatically decreasing mortality from diphtheria. Following this, Behring managed to develop another effective serum therapy against tetanus.[17,18]

Over the ensuring decades a number of other serum therapies were developed by other scientists, including one for pneumonia in the 1920s, and by the 1930s serum therapy was routinely being used for the treatment of many different infectious diseases, including erysipelas, scarlet fever, whooping cough, dysentery, measles, poliomyelitis, mumps, influenza meningitis and chickenpox.[19]

Serum therapy, however, carried the risk of side effects. Almost all patients manifested some form of adverse reaction to the treatment. These could range from the mild to very serious. The symptoms included fever, rashes and joint pains through to anaphylactic shock. Such problems had been observed by Behring as early as 1893. He and others believed they were caused by the fact that serum preparations contained proteins that were foreign to humans because they came from animals. Various attempts were made to improve the preparations of animal serum over the years which helped to reduce some of the reactions patients experienced. Nonetheless, problems continued.[20]

Not surprisingly the use of serum therapy fell out of favour following the introduction of sulphonamides in the 1930s and antibiotics in the 1940s. In part, this was because the new drugs were not only less toxic to patients, they were also easier to produce. Serum preparations had the major disadvantage that they required extensive characterisation and testing because of the variation between different batches. This made their preparation time-consuming and expensive. Moreover, the new therapies were easier to administer to patients. Serum preparations could only be given by intravenous injection, which required extensive skill on the part of the medical practitioner.[19]

4.8   A HESITANT START IN THERAPEUTICS

As soon as Mabs appeared, many clinicians enthusiastically began exploring their use for therapy. Indeed, many believed Mabs could provide the long-sought after 'magic bullet' envisioned by Ehrlich. Where they were seen as particularly exciting was for the treatment of cancer. One of the key advantages Mabs promised was that they could be used to specifically target cancer while leaving healthy cells alone. Work in this area, however, was not straightforward, hampered by the difficulty of finding an antigen on cancer cells that was not also present on other cells.[21]

In the end the first Mab that would make it to market would be in the field of organ transplantation and not cancer. In 1986 the

US Food and Drug Administration approved Orthoclone, a drug for the prevention of kidney rejection in transplant patients. Taking just seven years to develop, a much shorter time than the average eight to ten years for most drugs, the drug was hailed as a major milestone for transplant medicine. Expectations were running high that other Mab therapeutics would soon follow.

It quickly became apparent, however, that Mab drugs could cause many of the same side effects observed with serum therapy. Between 5 and 10% of patients on Orthoclone, for example, experienced significant adverse reactions, including fevers, anaphylactic shock and thromboses. It also carried the risk of severe infections and cancer.[22] Such problems were also observed in clinical trials of other Mab therapies.

One of the key difficulties was the drugs were made with murine antibodies, that is, they were derived from mice and rats. This meant they were regarded as foreign by the human immune system. As many as half of the patients treated with such antibodies experienced immune reactions. This not only threatened their health but led to the rapid destruction and clearance of the antibodies from the body, which undermined their therapeutic effect. Overall the antibodies lasted between 15 and 30 hours in humans. To get over this problem clinicians needed to inject high and frequent doses of the antibody into patients. A further problem was the fact that the antibodies were poor at recognising receptors on human cells.[23]

4.9   ENGINEERING HUMAN MONOCLONAL ANTIBODIES

From the late 1970s scientists, including Milstein, began looking for ways to produce human antibodies. This, however, was a major challenge. It was compounded by the fact that the human immune system is intrinsically tolerant of most human antigens and so does not produce many antibodies of use for therapy. An added complication is the fact that humans cannot be immunised and manipulated like laboratory animals. Immunising a human with a cancer antigen, for example, to raise antibodies against cancer for conversion into Mabs is risky and poses major ethical questions.

Faced with these obstacles genetic engineering offered a way forward. Milstein was one of the first people to advocate this

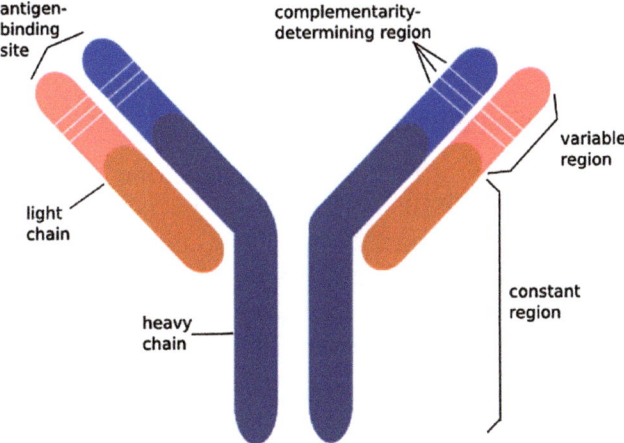

Figure 4.2   The basic structure of an antibody.
Source: L.V. Marks, 'A Healthcare Revolution in the Making: The Story of César Milstein and Monoclonal Antibodies', a digital exhibition, WhatIsBiotechnology.org/exhibitions/milstein.

approach. He realised it could provide a means to move away from merely immortalising naturally occurring antibodies in the way that he and Köhler had done and provide a way to design tailor-made Mabs. Those looking to adopt genetic engineering would be helped by the fact that antibodies possess a basic uniform modular scaffold with constant and variable regions (Figure 4.2). It was therefore possible to cut genes from a region of one antibody and paste them onto another antibody.

The idea of switching genes from one part of an antibody mirrored what the immune system does naturally. This process had first been established by Nobumichi Hozumi and Susumu Tonegawa in 1976.[24] Overall the body produces five different classes of antibodies, also known as immunoglobulins (Ig). Each class has a unique chemical structure and function. The two main classes are IgM and IgG. Whenever the immune system encounters an antigen the first class of antibody it produces is an IgM. Later on the immune system will covert these antibodies into IgG antibodies. It is these antibodies which provide long-term immunity. This is achieved by the rearrangement of genes between the variable and constant regions. Only three or four separate DNA segments are responsible for the vast array of antibodies the body produces. This opened up the possibility for scientists to combine segments of one antibody gene with the segments of another to create new antibodies.

4.9.1 Humanising Murine Antibodies

By 1984 three different teams of scientists, one led by Michael Neuberger at the LMB, the second by Sherrie Morrison at Stanford University and a third by Gabrielle Boulianne at Toronto University, had found a way of using genetic engineering to produce a new type of Mab more acceptable to the human immune system. Part mouse and part human, these Mabs were called 'chimeric' antibodies, nicknamed after chimera, the mythical creature made up of parts from multiple animals.[25–28]

Producing chimeric antibodies was a major undertaking because it was not possible to just pluck an appropriate DNA segment off the shelf. First, the scientists needed to identify the variable region of interest, create a library of DNA fragments taken from the region, and then clone the DNA of interest. Following this the DNA had to be put into a plasmid—an independent, self-replicating DNA molecule—and inserted into a myeloma cell to express the recombinant gene. This was a time-consuming and painstaking process. Once all this had been done the genes could be shuffled from the variable to the invariable regions of the antibody.[29]

In 1986, soon after the emergence of the first chimeric antibodies, Gregory Winter, a scientist based in the LMB, demonstrated the feasibility of decreasing the mouse component of the antibody to just 5%. Winter's technique involved the construction of a new antibody by grafting the antigen-binding loops, or complementarity-determining regions (CDRs; see Figure 4.2) from a mouse antibody onto a human one. Dubbed 'CDR grafting' this technique represented a major technical breakthrough in antibody engineering, generating what would come to be known as 'humanised' Mabs.[30]

4.9.2 Fully Human Antibodies

Not content with humanising rodent antibodies, in 1987 Winter began looking for a way to create an artificial immune system that could produce fully human Mabs. His aim was to build a large library of human antibody fragments, which could be screened for those which bound to desired antigens. Once identified the fragments could be used to engineer a fully human Mab.

4.9.2.1 Phage Display Technology. By 1990 Winter had fulfilled his mission. He and his team had found a way of generating enormous diverse libraries of antibody fragments. This they achieved with the help of the polymerase chain reaction (PCR), a laboratory technique, first developed in 1983, which created millions of copies of very small samples of DNA. PCR provided the means to amplify and clone genes directly from the lymphocytes of an animal. Once copied and cloned the genes were then inserted into phages, viruses that infect bacteria, to engineer phages that displayed antibody fragments on their coat surface.[31,32] This was built on a technique originally developed for the display of peptides in 1985.

Once made, the phages displaying the antibody fragments were assembled into a library ready for screening. This was done by adding the phage-display library to the wells of a microtitre plate that contained immobilised targets of interest. These targets could be, for example, an antigen from a cancer cell or a molecule known to cause inflammation. The plate was then incubated to allow the phages to bind to the target and then washed to flush away any non-binding phages. Any phages that remained attached to the wells were removed and inserted into bacteria for replication. These steps were repeated until only phage-displaying antibody fragments for the target remained. One of the advantages with the new platform was many different antibody fragments could be screened at any one time. Once the screening process was completed, the gene coding for the specific antibody was isolated and purified from the phage which could be stored, amplified or processed in other ways ready for use to produce a Mab.

Called antibody phage display technology, the new method marked a major turning point in the engineering of Mabs. Now scientists were no longer reliant on an animal or human's immune system with all their limitations, or having to go through the time-consuming and laborious process of making a hybrid cell. The platform also provided much greater scope to tailor the specificity and affinity of the Mab and allowed for the creation of synthetic Mabs with less chance of provoking an immune response. In addition, the process was quick, taking approximately two weeks to generate a Mab of interest, and could be automated.[33–36]

4.9.2.2 *Transgenic Animals.*

In parallel with the development of the phage display technology, another avenue was opening up for the production of human antibodies. This centred on the genetic modification of mice to generate human antibodies. Scientists had been successfully making transgenic mice since the mid-1970s. This involved the introduction of foreign DNA segments into the germ-line of early embryos. The objective was to either inactivate a mouse's own genes or to introduce new genes.

One of the first pioneers in the field was Michael Neuberger at the LMB. He started work on the project in 1986. This he did with colleagues at the Babraham Institute, Cambridge, a pioneering centre in animal genetic engineering. By the late 1980s two American teams, one based in GenPharm and another at Cell Gensys, two Californian start-up companies, had also become active in the area. All three had achieved their objective by 1993 using broadly similar techniques. The key challenge they had faced was making sure that the genes they inserted into the mouse assembled successfully and interacted with the necessary signalling components in the mouse's lymphocytes so that when it was immunised it produced human antibodies to the desired antigen.[37]

A significant advantage with the transgenic mice was that they enabled the production of fully human Mabs with enhanced affinity for a target without spending hours genetically engineering a particular antibody molecule. Everything could now be achieved in the mouse. Once a transgenic mouse was immunised it produced a target-specific human Mab without any subsequent manipulation. This meant that Mab drugs could be developed much faster than before.

4.10   THE CHANGING COMPOSITION OF MABS

All the advances in Mab engineering dramatically changed the Mab molecule from the time of Milstein and Köhler's invention. As Figure 4.3 and Table 4.1 highlight, such work helped reduce the mouse component to almost nothing, making them much less likely to provoke negative responses from the human immune system. This has been reflected by the type of Mab drugs that have been approved over the years. While it was a murine

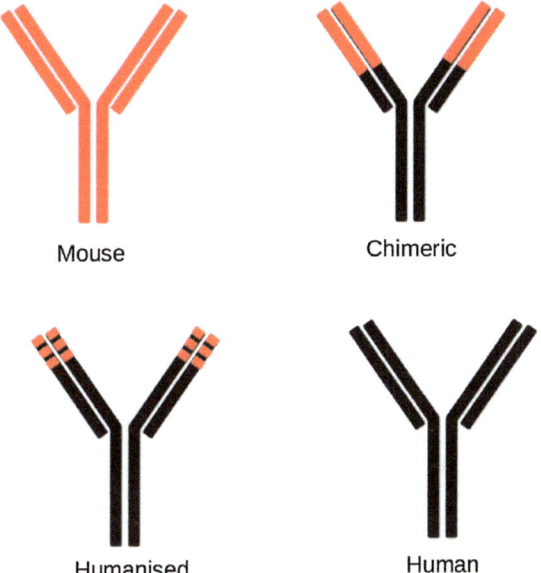

Figure 4.3   The different protein components of engineered Mabs.

Table 4.1   Table showing percentages of mouse and human protein components in different forms of Mabs. Data taken from ref. 38.[a]

Form of Mab	Mouse protein	Human protein
Murine	100%	0%
Chimeric	65%	35%
Humanised	95%	5%
'Fully human'	0%	100%

[a]W. R. Gombotz and S. J. Shire, Introduction, in *Current Trends in Monoclonal Antibody Development and Manufacturing*, New York, 2009, p. 3.

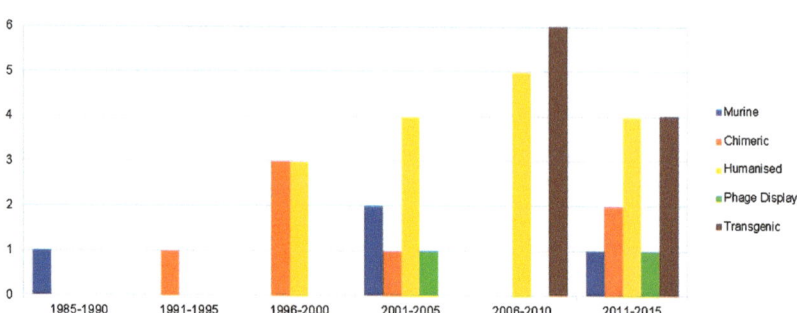

Figure 4.4   Mabs by type approved for therapeutic use from 1985–2015.

Mab that was first approved for the market, by 1993 the first chimeric had appeared on the market. The first humanised Mab, made with CDR grafting, appeared on the scene in 1997 and the first fully human Mab drug, developed with phage display, in 2002. Four years later the first human Mab produced using a transgenic mouse was approved. Figure 4.4 provides a summary of the different Mabs approved for the market between 1985 and 2015.

Scientists did not stop at reducing the mouse protein in Mabs. They soon began to explore the possibility of using just the active portion of antibodies for treatment. Started by Winter in 2002, these efforts took time to succeed. Much of the work was driven by two factors. Firstly, it could help reduce the amount of antibodies that needed to be given to a patient. Up to then large quantities of Mabs had been needed to achieve clinical efficacy. Such high volumes posed significant issues in terms of production, which added to cost of the drug. Secondly, it was believed that such fragments might be better at penetrating tissues than a conventional Mab, which is a very large molecule. In addition it was hoped such fragments might offer better targeting of pathogenic molecules, which remained largely inaccessible to full-sized Mabs due to the narrow cavities on their surface antigens. The first Mab fragment was approved for market in 2007.[39]

4.11   DISEASE CONDITIONS

The improvements made to Mabs by the different genetic engineering techniques reawakened interest in their therapeutic potential. Early enthusiasm for Mabs as drugs had already begun to fade by the early 1990s. This was in part because of the side effects witnessed with the murine Mabs but also because in 1992 the sector had experienced a major shock. That year a widely acclaimed drug, Centoxin, against all expectations had failed to win US approval for the treatment of septic shock. The event had not only nearly bankrupted Centocor, the biotechnology company responsible for Centoxin's development, but greatly dampened investment in the field overall. In 1994, however, against all predictions, Centocor managed to bounce back by gaining approval for ReoPro, another Mab drug it had had in its pipeline. Used to prevent ischemic complications in patients undergoing coronary angioplasty, a common procedure to unblock coronary arteries, ReoPro was a chimeric Mab.

The approval of ReoPro not only saved Centocor, but placed Mabs firmly on the therapeutic map. As of November 2014 47 Mab drugs had been approved for in the USA and/or Europe since 1986, 18 of which had achieved annual sales over US$1 billion, with six of them being more than US$6 billion. One product, Humira, had achieved the highest figure ever recorded for a biopharmaceutical product—US$11 billion. Moreover, hundreds more Mab drugs were in the pipeline.[40]

Mab drugs have been developed for multiple conditions. Figure 4.5 provides a summary of the variety of diseases for which Mab drugs were approved by the FDA between 1993 and August 2016.

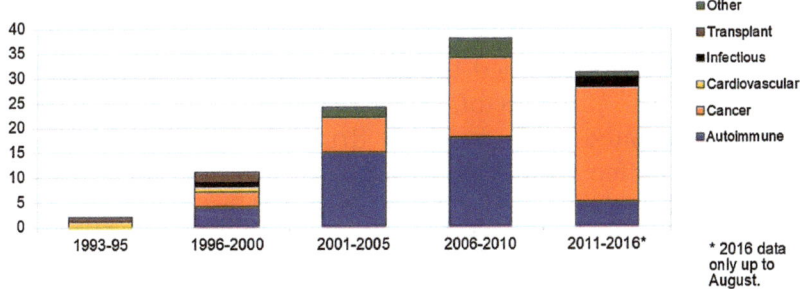

Figure 4.5   Number of Mab drugs approved by FDA by different categories 1993–Aug 2016 (includes first and supplementary approvals). Data taken from ref. 41 and FDA, *Novel Drug Approvals*, 2011–2016.

4.11.1   Auto-immune Disorders

As can be seen from Figure 4.5 auto-immune disorders were one of the first classes of conditions where Mabs made their mark, forming the bulk of the Mab drugs approved between 2001 and 2005. In part, this reflected the fact that early on Mabs played a pivotal role both as a laboratory tool for investigating the pathogenesis of auto-immune disorders and as a device to treat them.

Where Mabs have proven particularly helpful has been in the management of rheumatoid arthritis, Crohn's disease and multiple sclerosis. Prior to the arrival of Mab drugs, most of these conditions had been treated with steroidal drugs. These drugs, however, take time to take effect and are highly toxic. Moreover, they can only slow down the progression of the disease. In contrast, Mab-based drugs are designed to modify the underlying mechanism that triggers the disease. In this way Mabs have helped transform the treatment of such diseases from merely ameliorating the symptoms to blocking their cause. Furthermore, the Mab drugs work quickly in patients and provide sustained clinical improvements.[42,43]

4.11.2   Cancer

Strikingly, Mabs for auto-immune disorders largely overshadowed the approval of Mab drugs for cancer in the early years (Figure 4.5). Nonetheless, Mabs quickly proved to be powerful tools for identifying and targeting different antigens on tumours. Although these applications did not quickly translate into successful therapeutics for cancer, in more recent years Mab therapeutics have transformed the care of cancer patients, who no longer face the prospect of losing hair and other serious side effects associated with previous drugs, which had a broad spectrum of effects and high toxicity.

One of the advantages of Mabs is that they can be given as maintenance therapies. This has reshaped our perception of some cancers from what was once seen as fatal to a chronic condition. Mabs have also opened up the possibility of matching therapies with particular tumour antigens in individual patients. This has allowed for a greater degree of personalisation in the management of cancer than was possible in the past. Indeed, Mab therapeutics are expected to become an increasingly important component of personalised cancer therapies. They are also critical components of immunotherapy for cancer, a treatment mode outlined in Chapter 5 that has been growing for cancer in recent years. The rise in the importance of Mabs for cancer is reflected in Figure 4.5, which shows that between 2011 and 2016 the majority of Mab therapeutics approved were for cancer treatment.

4.11.3 Other Conditions

Over time, Mab drugs have been approved for a myriad of conditions beyond auto-immune disorders and cancer. For example, by 2003 the first Mab therapeutic had been introduced on to the market for asthma and in 2006 another had been introduced for an eye disorder (neovascular age-related macular degeneration). In 2010 the first Mab drug was approved for osteoporosis. Mab drugs have also begun to be investigated for the treatment of infectious diseases. To-date only two anti-infective Mab drugs have been licensed; the first was approved in 1998 for respiratory syncytial virus in high-risk premature babies and the second in 2012 for the treatment of inhalation anthrax. The slow progress of Mab therapeutics for infectious disease can, in part, be attributed to the large arsenal of other anti-infective drugs, such as vaccines and antibiotics. Researchers are also now very active in looking at Mabs as a potential for the treatment of Alzheimer's and even mental disorders like depression and schizophrenia which are increasingly thought to be linked to inflammation.[44]

4.12 THE SUCCESS OF MABS

It is now over 40 years since Mabs first arrived on the scene. In that time Mabs have radically reshaped medicine and spawned a whole new industry. Some idea of how much the monoclonal market has grown can be seen from Table 4.2. The first products to reach the market were diagnostics. These were to be followed by therapeutics, which gathered momentum from the late 1990s. Between 2008 and 2013 the global sale revenues for Mab drugs increased from approximately $39 billion to nearly $75 billion.

Table 4.2   Global sales (billions) of Mab therapeutic and diagnostic antibodies from 2005–2012. Data taken from ref. 45.[a]

Technology type	2005	2012	Average annual growth rate
Therapeutic	13	47	14%
Diagnostic	5.7	9	7%
Total	19	56	13%

[a]Source: BCC Research, 2012.

Mab drugs comprised nearly 50 per cent of the total sales of all biopharmaceutical drugs. As of 2015 47 Mab drugs had been approved for the US and European market for several diseases and it was anticipated that a total of 70 Mab products would be on the market by 2020. Six of the drugs were each generating more than $6 billion a year. One of the drugs, adalimumab, generated an annual revenue of nearly $11 billion. This was the highest annual sales figure documented to date for a biopharmaceutical product. First approved for treating rheumatoid arthritis in 2002, adilmumab was created with the help of Winter's pioneering techniques and was the world's first fully human antibody.[40]

4.13   THE COST DILEMMA

Despite their success, Mab drugs have not provided a total panacea. Indeed, they pose significant risks in terms of side effects. Furthermore, they come with a high price tag. One drug, eculizumab, approved to treat a rare blood disorder cost $440 500 per patient in the USA in 2016. It is one of the most expensive drugs in the world. Not all Mab drugs come at such a high price. Nonetheless, even trastuzumab, a common treatment for breast cancer cost $64 000 per patient in the USA in 2015. The high price attached to Mab drugs has placed them at the centre of debates about the rising cost of healthcare and how universally they should be provided.

In part, the high cost of Mabs reflects the high pricing point generally associated with early innovative treatments and the fact that such drugs are still under patent. Like other drugs, Mab drugs have patents that normally run for 20 years. On average at least half of the time of the patent will be spent proving the drug's efficacy and safety. Once on the market a manufacturer can sell the drug at a price to maximise profits. This enables them to recoup some of the research and development costs they expend in getting to the drug to market.

The total amount companies spend before a drug is approved has been fiercely contested. Figures analysed from 12 leading pharmaceutical companies for the years 1997 to 2011 suggest that pharmaceutical companies spent on average US$802 billion to gain approval for just 139 drugs (US$5.8 billion per drug). A key factor in the development costs of a drug are the clinical trials. This, in part, reflects the complex regulation involved, which has become more burdensome in recent years and resulted in an increasing demand for larger and more complex trials. Research undertaken by researchers at Tufts University showed that between 1999 and 2005 the average length of a clinical trial increased by 70%, the average number of routine procedures by 65% and the average clinical trial staff work burden by 65%. In addition to this, increasingly stringent enrolment criteria and trial protocols resulted in 21% fewer volunteers being admitted into trials and 30% dropping out before completion of the tests. The most costly aspect of clinical trials are those undertaken at the phase III stage, which is used to confirm the effectiveness of a drug and monitor its mid- to long-term side-effects in large groups of patients. Today phase III trials represent about 40% of the research and development costs incurred by pharmaceutical companies.[46]

Many drugs become much less expensive once their patents expire. It is difficult to know whether this will be the case for Mab therapeutics. US and European regulatory authorities have recently set out their approval process to facilitate the marketing of biosimilar (generic) Mab therapeutics. In June 2013 the European Medicines Agency recommended the marketing authorisation for the first Mab biosimilar, which contains the same known active substance as infliximab which was approved for the treatment of auto-immune diseases such as rheumatoid arthritis, Crohn's disease, ulcerative colitis, ankylosing spondylitis, psoriatic arthritis and psoriasis. Many more Mab biosimilars are in the pipeline. The bulk of these are for cancer, and the remainder for autoimmune/inflammation, metabolic, infectious disease and central nervous system indications.[47]

Analysts generally believe the cost and complexity of manufacturing and testing biosimilars will keep the cost high. The price of biosimilars is predicted to be only between 20 and 30% less than that of patented therapeutics. This is much lower than small-molecule drugs which cost between 80 and 90% less once off patent. Overall, the barriers to the development of biosimilar Mabs are likely to be much greater compared with those of generic pharmaceuticals. Any inadvertent chemical modifications can affect the performance of Mabs in humans and their safety profile, and it is difficult to predict their immunogenicity. As a result, regulatory authorities require clinical trials to demonstrate equivalence for efficacy, unlike the case with small-molecule generics. To have sufficient power to be capable of detecting differences from the original drug, such trials often have to be larger than before. The biosimilar Mab granted European marketing authorisation, underwent clinical testing in 874 patients in 20 countries across 115 sites. The cost of clinical testing in itself can be prohibitive. Biosimilar drugs also need state-of-the-art manufacturing technology, which is both expensive and cumbersome. For the time-being, therefore, development costs will remain high, thereby inhibiting many companies from undertaking such a venture.[48–51]

One of the factors boosting the cost of Mab therapeutics is the fact that they have a relatively low potency. In contrast to most other drugs, including other biological therapeutics, which are mostly given in milligram quantities, Mab drugs generally need to be administered in grams. The recommended dose for adalimumab, for example, is 40 milligrams every two weeks, totalling over 1 gram per year. Large volumes of Mabs are therefore needed to meet market demand, ranging from tens to hundreds of kilograms per year.[50,52] Some of the difficulties faced in manufacturing these quantities of Mabs are outlined in Chapter 2 by Alldread and Birch.

4.14 CONCLUSION

Despite their high cost, the development of Mabs has gone some way towards realising Paul Ehrlich's original vision that one day antibodies could provide the 'magic bullet' for treating disease.

While generally receiving less public fanfare than many other forms of biotechnology, such as genetic engineering and stem cells, Mabs have transformed the healthcare landscape. Quietly they have helped unlock the power to probe many different pathways of disease within the laboratory and radically improved the speed and accuracy of diagnostics. In addition, they have opened up new avenues for treatment, helping to bring relief to millions of suffers across the world.

REFERENCES

1. L. V. Marks, *The Lock and Key of Medicine: Monoclonal Antibodies and the Transformation of Healthcare*, Yale University Press, 2015.
2. E. Behring, *Dtsch. Med. Wochenschr.*, 1890, **16**, 1145–1148.
3. E. Behring and S. Kitasato, *Dtsch. Med. Wochenschr.*, 1890, **16**, 1145–1148.
4. *Milestones in Microbiology: 1556 to 1940*, ed. T. S. D. Brock, ASM Press, 1998, pp. 138–141.
5. P. Ehrlich, *Dtsch. Med. Wochenschr.*, 1891, **17**(976–9), 1218–1219.
6. P. Ehrlich, *Ned. Tijdschr. Geneeskd.*, 1909, **5**, 273–290.
7. H. P. Vollmers and S. Brändlein, *Histol. Histopathol.*, 2005, **20**, 927–937.
8. A. M. Silverstein, *Cell. Immunol.*, 1999, **194**, 213–221.
9. C. R. Prüll, *Med. Hist.*, 2003, **47**, 332–356.
10. L. V. Marks, A heathcare revolution in the making: The Story of César Milstein and monoclonal antibodies, Section 4, http://www.whatisbiotechnology.org/exhibitions/milstein/monoclonals.
11. G. Köhler and C. Milstein, *Nature*, 1975, **256**(5517), 495–497.
12. P. Keating and A. Cambrosio, *Biomedical Platforms: Realigning the Normal and the Pathological in Late Twentieth Century Medicine*, MIT Press, 2003, p. 172l.
13. L. V. Marks, *The Lock and Key of Medicine: Monoclonal Antibodies and the Transformation of Healthcare*, Yale University Press, 2015, ch. 3.
14. L. C. Goldstein and M. R. Tam, *Clin. Lab. Med.*, 1985, **5**(3), 75–78.
15. S. Chen and S. C. Silverstein, *FASEB J.*, 1993, 7, 1426–1432.
16. M. G. Scott, *Trends Biotechnol.*, 1985, **3**(7), 170–175.

17. D. S. Linton, *Emil von Behring: Infectious Disease, Immunology, Serum Therapy*, American Philosophical Society, 2005.
18. A. C. Hüntelmann, *Dynamis*, 2007, **27**, 107–131.
19. A. Cassadevall and M. D. Scharff, *Clin. Infect. Dis.*, 1995, **21**, 150–161.
20. L. V. Marks, *The Lock and Key of Medicine: Monoclonal Antibodies and the Transformation of Healthcare*, Yale University Press, 2015, pp. 10–11.
21. L. V. Marks, *The Lock and Key of Medicine: Monoclonal Antibodies and the Transformation of Healthcare*, Yale University Press, 2015, ch. 5.
22. C. Sgro, *Toxicology*, 1995, **105**(1), 23–29.
23. A. F. LoBuglio, R. H. Wheeler, J. Trang, A. Haynes, K. Rogers, E. B. Harvey, L. Sun, J. Grayeb and M. B. Khazaeli, *Proc. Natl. Acad. Sci. U. S. A.*, 1989, **86**, 4220–4224.
24. N. Hozumi and S. Tonegawa, *Proc. Natl. Acad. Sci. U. S. A.*, 1976, **73**(10), 3628–3632.
25. M. S. Neuberger, G. T. Williams and R. O. Fox, *Nature*, 1984, **312**(5995), 604–608.
26. M. S. Neuberger, G. T. Williams, E. B. Mitchell, S. S. Jouhal, J. G. Flanagan and T. H. Rabbits, *Nature*, 1985, **314**, 268–270.
27. L. Morrison, M. J. Johnson, L. A. Herzenberg and V. T. Oi, *Proc. Natl. Acad. Sci. U. S. A.*, 1984, **81**, 6851–6855.
28. G. Boulianne, N. Hozumi and M. J. Shulman, *Nature*, 1984, **312**, 643–646.
29. L. V. Marks, *The Lock and Key of Medicine: Monoclonal Antibodies and the Transformation of Healthcare*, Yale University Press, 2015, pp. 161–163.
30. G. P. Winter, *Philos. Trans. R. Soc. London, Ser. B*, 1989, **324**, 537–547.
31. R. Orlandi, D. H. Gussow, P. T. Jones and G. Winter, *Proc. Natl. Acad. Sci. U. S. A.*, 1989, **86**, 3833–3837.
32. J. McCafferty, A. D. Griffiths, G. Winter and D. J. Chiswell, *Nature*, 1990, **348**, 552–554.
33. H. R. Hoogenboom and G. Winter, *J. Mol. Biol.*, 1992, **227**, 381–388.
34. A. Nissim, H. R. Hoogenboom, I. M. Tomlinson, G. Flynn, C. Midgley, D. Lane and G. Winter, *EMBO J.*, 1994, **13**(3), 692–698.
35. H. R. Hoogenboom, *Tibtech*, 1997, **15**, 62–70.

36. S. J. Russell, M. B. Llewelyn and R. E. Hawkins, *Br. Med. J.*, 1992, **304**, 585–586.
37. L. V. Marks, *The Lock and Key of Medicine: Monoclonal Antibodies and the Transformation of Healthcare*, Yale University Press, 2015, pp. 168–173.
38. W. R. Gombotz and S. J. Shire, Introduction, in *Current Trends*, ed. S. J. Shire, *et al.*, American Association of Pharmaceutical Sciences, 2010.
39. A. L. Nelson, *mAbs*, 2010, **2**(1), 77–83.
40. D. M. Ecker, S. D. Jones and H. L. Levine, *mAbs*, 2015, **7**(1), 9–14.
41. K. Stein, FDA-approved monoclonal antibody products, 2010, unpublished paper.
42. S. Abramson, Expected outcomes in rheumatoid arthritis: An historical perspective, Medscape Education, http://www.medscape.org/viewarticle/464118_2. Accessed Feb 2017.
43. P. Emery, *Br. Med. J.*, 2006, **332**, 152–155.
44. J. Gallaghher, R. Buchanan and A. Luck-Baker, Depression: A revolution in treatment, *BBC Health*, 24 August 2016, http://www.bbc.co.uk/news/health-37166293. Accessed Feb 2017.
45. BCC Research, 2012, http://www.businesswire.com/news/home/20071212005806/en/Pharmaceutical-Market-.
46. A. S. A. Roy, Stifling new cures: The true cost of lengthy clinical drug trials, *Manhattan Institute for Policy Research*, April 2012, https://www.manhattan-institute.org/html/stifling-new-cures-true-cost-lengthy-clinical-drug-trials-6013.html. Accessed Feb 2017.
47. F. Barkalow, Biosimilar monoclonal antibodies in the pipeline: Major players and strategies, http://www.biotechduediligence.com/uploads/6/3/6/7/6367956/biosimilars_citeline.pdf. Accessed Feb 2017.
48. FDA, News release, FDA issues draft guidance on biosimilar product development, https://www.fda.gov/NewsEvents/Newsroom/PressAnnouncements/ucm291232.htm. Accessed Feb 2017.
49. M. Becker, Monoclonal antibody companies command premiums, July 2010, http://seekingalpha.com/article/214040-monoclonal-antibody-companies-command-premiums. Accessed Feb 2017.
50. G. C. Fanneau La Horie, Making biologic drugs more affordable, *Drug Discovery and Development*, 10 July 2010,

http://www.dddmag.com/article/2010/07/making-biologic-drugs-more-affordable. Accessed Feb 2017.

51. C. K. Schneider and U. Kalinke, *Nat. Biotechnol.*, 2008, **26**(9), 985–989.

52. S. S. Farid, *J. Chromatogr. B: Anal. Technol. Biomed. Life Sci.*, 2007, **848**, 8–18.

CHAPTER 5

The Changing Fortune of Cancer Immunotherapy[†]

LARA V. MARKS

University College London, UK
Email: l.marks@ucl.ac.uk

5.1 INTRODUCTION

In 2013 the editors of the prestigious American journal, *Science*, nominated cancer immunotherapy the 'Breakthrough of the Year'. They argued that 'clinical trials ... cemented its potential in patients and swayed even the sceptics. The field hums with stories of lives extended: the woman with a grapefruit-size tumour in her lung from melanoma, alive and healthy 13 years later; the 6-year-old near death from leukaemia, now in third grade and in remission; the man with metastatic kidney cancer whose disease continued fading away even after treatment stopped.'[1]

Immunotherapy is designed to boost the body's immune system to combat cancer. Such therapy takes two key forms.

[†]I am grateful to Don Drakeman, Nils Lonberg, Ilana Löwy and Nils Graber for comments on earlier drafts of this chapter.

Engineering Health: How Biotechnology Changed Medicine
Edited by Lara V. Marks
© The Royal Society of Chemistry 2018
Published by the Royal Society of Chemistry, www.rsc.org

The first, known as active immunotherapy, aims to activate the body's own, endogenous, immune system to fight cancer cells. The second, known as passive immunotherapy, directly targets cancer cells. This is done by injecting immune cells capable of attacking the cancer, which have been produced in the laboratory, into the patient. A number of approaches are used for immunotherapy. These include re-activating a switch in immune cells that tumour cells turn off to prevent their own destruction, tagging cancer cells for their elimination by immune cells, or genetically modifying a patient's own T cells, the foot soldiers of the immune system to destroy the cancer.

In 2017, for the second year in succession, the American Society of Clinical Oncology, cited immunotherapy as one of the most important developments in the medical treatment of cancer in recent times. This enthusiasm was fuelled by a new class of drugs known as immune checkpoint inhibitors. These are designed to block the biological pathway cancer cells use to prevent their own destruction by hiding themselves from the immune system. By March 2017 five such drugs had been approved for the US market and several were in late-stage clinical testing. What is most promising is that the drugs are prolonging the survival of patients with advanced cancer for whom there were previously no other options.

The idea of using the immune system to fight cancer has a long history. Yet, as this chapter shows, it took many years for the concept to be fully recognised and treatment to be rolled out on a significant scale.

5.2   EARLY CANCER VACCINE ATTEMPTS

As early as the eighteenth century, a number of physicians observed that some cancer patients went into remission following a fever caused by an infection. One of the first to record this phenomenon was the French physician Antoine Diedier. In 1725, he noted that syphilis sufferers developed very few malignant tumours.[2] More than a century later, in the 1860s, two German physicians, W. Busch and Friedrich Fehleisen, independently noticed the shrinking of tumours in patients who accidentally contracted erysipelas, a skin infection. Based on this observation and the success of the smallpox vaccine, in

1868 Busch deliberately infected a nineteen-year-old woman with erysipelas in an attempt to eliminate a sarcoma which could not be removed surgically. The woman did not survive, but her tumour shrank to half its size.[3] Following this, in 1882, Fehleisen demonstrated that erysipelas could destroy transplanted tumours in animals, and went on to inject pure cultures of erysipelas into seven patients with inoperable cancer. Three of these patients, two with breast cancer and one with sarcoma, experienced a significant decrease in the size of their tumours.[4]

Both Busch and Fehleisen conducted their experiments without any knowledge of the streptococcal organism that causes erysipelas. This was only identified by Fehleisen in 1882 after he had treated his patients. Six years later, another German physician, P. Bruns, claimed he had cured three out of five patients with sarcoma by administering injections of *Streptococcus pyogenes*, the bacterial organism that causes erysipelas.[5]

Soon after the German trials, William Coley, a surgeon based at Memorial Hospital in New York began combing his hospital's medical records for clues to help him improve cancer treatment. He was prompted to do this by the loss of his first cancer patient, a 17 year-old woman whose arm he had amputated in a desperate attempt to arrest the spread of aggressive sarcoma. One of the records was particularly intriguing. It was of a man who had survived four episodes of recurrent inoperable sarcoma of the neck following a severe erysipelas infection. Coley subsequently found 38 similar cases in the medical literature. Based on these findings, in 1891 he decided to inject a culture of streptococcal bacteria into a 35 year-old man with a tumour 'the size of small hen's egg' on his neck. The man had only weeks to live. To Coley's delight the patient experienced a complete remission. By 1893 Coley had used the same approach in a further nine sarcoma patients, a number of whose cancers improved.[6,7]

Over the next few years Coley continued to perfect his technique. This he pursued on the premise that cancer was associated with a pathogen which could be combated with a vaccine approach. His method was not without its problems. Critically, it did not always induce an infection and when successful it could be difficult to control. In order to resolve these issues Coley began experimenting with different vaccine formulations. These contained a combination of heat-killed streptococcus with *Serratia marcesecens*, another bacterial agent.[8,9]

By the time of his death in 1936, Coley and his colleagues had treated nearly 1000 patients with 13 different vaccine formulations. One of the vaccines they reported effected a cure in 60 out of 210 patients with terminal soft-tissue sarcoma, many of whom were still living ten years after treatment. Coley interpreted the results as the treatment stimulating the body's resistance to cancer. He did not, however, make any attempt to understand what mechanism was involved. The anti-tumour effects of the vaccine, he assumed, were due to the bacterial origins of the cancer. Coley's treatment had been largely abandoned by the 1940s. What hindered its continuance was the fact that Coley had never systematically tested his method. Nor had he properly codified its application. Many researchers thus struggled to replicate Coley's success. In addition, the treatment had unpleasant side effects, including high fever and pain.[8,9]

Erysipelas was not the only type of bacterial infection explored for defeating cancer. In 1929 Raymond Pearl, a biologist at Johns Hopkins University, observed, from an investigation of 1632 patient autopsies, that patients with tuberculosis appeared to have a much lower incidence of cancer. Based on this, he persuaded one of his clinical colleagues to test a vaccine against tuberculosis in cancer patients.[10] This had been developed in 1921. It was known as the Bacille Calmette–Guerin (BCG) vaccine. While the vaccine was only given to a handful of patients, the method appeared effective.[11] Somewhat later, in 1935, I Holmgren, a clinician based in Stockholm, reported successfully using the vaccine to treat 28 patients with stomach cancer.[12,13] The following year Sol Roy Rosenthal, the director of the Institute for Tuberculosis Research at the University of Illinois, found that the vaccine was capable of stimulating the macrophage system, which was known to play an important role in the prevention of cancer.[14,15]

Despite its promise, most trials with the BCG vaccine for cancer had been discontinued by the late 1930s. This followed a disastrous event in 1930. That year a large number of German babies contracted tuberculosis when they received injections of a vaccine preparation contaminated by a virulent strain of tuberculosis. It had been produced by a laboratory in Lübeck, a town in Northern Germany. Over 70 children died. The tragedy shattered confidence in the vaccine world-wide, not only for preventing tuberculosis but also for treating cancer.[16,17]

5.3 SERUM THERAPY

Vaccination was not the only type of early treatment tried against cancer. Patients were also administered injections of blood serum. This approach, known as serum therapy, followed a number of discoveries made in 1891. The first was made by Hans Buchner, a German bacteriologist, who noticed a soluble component in blood capable of killing bacteria.[18] Soon after this Emil von Behring, another German physician, and Kitasato Shibasaburō, a Japanese physician, jointly demonstrated that serum taken from animals that had survived diphtheria or tetanus could immunise uninfected animals and cure those already infected with the two diseases.[19] Shortly thereafter, the German physician Paul Ehrlich isolated a substance in serum which provided immunity against plant poisons. He called the substance an antibody.[20] In 1895 another protective molecule was detected in blood by Jules Bordet, a Belgian immunologist and microbiologist. Subsequently called 'complement', the substance was found to act as an accessory to the antibodies in destroying bacteria.[21]

The new understanding about the power of serum was soon put to therapeutic use. By 1893 Behring and Ehrlich had successfully pioneered a therapy for diphtheria using serum from horses immunised against the disease. Two years later Jules Héricourt and Charles Richet, two French physicians, reported positive responses for two cancer patients who had received injections of serum obtained from a donkey and two dogs that had been immunised previously with an extract of a human osteo-sarcoma tumour. Over the next two years they continued to use the same method in another 50 cancer cases, with similarly encouraging results. Other researchers would continue to investigate the use of serum to combat cancer into the early twentieth century but with varying success.[22]

5.4 SHIFTING ATTITUDES TO THE IMMUNE SYSTEM AND CANCER

Despite their potential, vaccines and serum therapy had largely fallen by the wayside for the treatment of cancer by the 1930s. In part, this was due to the rise of radiotherapy and chemotherapy, which were easier to use. Their abandonment also reflected a more general scepticism within the scientific community about the extent to which the immune system could recognise and destroy malignant tumours.[9] For many years immunologists wondered how the immune system could distinguish between tumour cells and those from healthy 'self' tissue, especially as cancer cells were able to transform themselves into a semblance of normal cells.[23–25]

During the 1950s a new avenue opened up for understanding the immune system and cancer through the development of new inbred strains of laboratory animals. A key breakthrough happened in 1957. That year Richmond Prehn and Joan Main, based at the US Public Health Service Hospital in Seattle, provided the first conclusive evidence that tumours carried specific markers, known as antigens, which the immune system could recognise and attack. This they demonstrated by chemically inducing tumours in inbred mice.[26] Two years later, Lloyd Old and colleagues at the Memorial Sloan Kettering Cancer Center, New York, demonstrated that the BCG vaccine could provoke an anti-tumour immune response in mice. They observed that the vaccine activated macrophages, a type of immune cell, which inhibited and destroyed the tumour cells.[27]

Evidence collected from the new animal studies helped lay the foundation for a new concept called 'immune surveillance'. This was developed by Frank MacFarlane Burnet, an Australian immunologist, in the late 1950s. Burnet proposed the hypothesis that the immune system regularly screened and protected the body against tumours, and that cancer only developed when the immune system acquired a tolerance to the cancer cells, thus allowing the cells to escape destruction and proliferate. Burnet went a step further to suggest that one way to treat cancer might be to increase the immune system's sensitivity to minor deviations from the body's own cells.[28,29]

Burnet's hypotheses reinvigorated research into immuno-therapy for cancer. In 1975, however, the field suffered a set-back as a result of the work of Osias Stutman and colleagues at the Memorial Sloan-Kettering Cancer Center. To everyone's surprise they had found that the incidence of tumours appeared to be the same in genetically bred laboratory mice with an inhibited im-mune system as in their wild cousins. The results threw into question whether the immune system could combat cancer.[24,30]

Interest in immunotherapy was soon reawakened by the work of Aline Van Pel and Thierry Boon at the Sloan-Kettering Cancer Center in 1982. They demonstrated, through mice experiments, that spontaneous tumours carried specific antigens but that these were often too weak to stimulate an effective immune response. In addition they showed that it was possible to enhance the im-mune response of the mice to the tumours by injecting tumour cells that had been genetically modified to increase their anti-genicity.[31] A few years later, Robert Schreiber and colleagues at Washington University showed that a new type of genetically modified mice, known as 'knockout mice', with totally inactivated immune systems had a very high incidence of tumours. This in-dicated that the mice studied by Stutman, called 'nude' mice, had not been as immunodeficient as previously assumed.[24,32]

5.5 INTERFERON: A NEW WEAPON EMERGES

As theories about the immune system and cancer twisted and turned, new tools emerged for testing the efficacy of immuno-therapy at a clinical level. One of the first was the discovery of interferons. This is a group of proteins secreted by the immune system. Interferons are part of a wider class of substances, called cytokines, which carry signals between cells. The first interferon was described in 1957 by Alick Isaacs and Jean Lindermann based at the National Institute of Medical Research, London. They named it interferon because of its ability to 'interfere' with viral growth and protect cells from viral infections.

Five years later, in 1962, a group of researchers, led by Kurt Paucker at the Children's Hospital in Philadelphia, noticed that interferon could also slow down the growth of certain tumours. By the late 1960s, Ion Gresser, an American virologist working at Institute Gustave Roussy in France, had further found that it could suppress virus-induced tumours in mice and stimulate lymphocytes to attack the tumours.[33]

Interferon immediately sparked excitement as a means to treat cancer. Trials with the substance were initially severely limited because it was difficult to produce. The first cancer patients received interferon in 1963. It was given to 11 patients with acute myeloid leukaemia, one of whom experienced some improvement. In 1969 Kari Cantell, a Finnish virologist, developed a method for the large-scale production of interferon. However, his preparation was crude: it contained only 1% interferon. One of the first clinicians to test Cantell's interferon was Hans Strander at Karolinska Hospital in Sweden, who started testing it in patients in 1971. By 1977 Strander had treated approximately 70 patients with osteogenic sarcoma, a highly aggressive cancer, with varying results. He had combined the use of interferon with other treatments.[34]

Until the early 1980s interferon was produced by purifying the substance secreted from human cells such as leukocytes (white blood cells) or fibroblasts (cell found in connective tissue) in test tubes. This all changed following the development of recombinant DNA and monoclonal antibodies, which provided the means to genetically engineer large quantities of interferon in bacteria and with much greater purity (see Chapter 2).[35] The mass production of interferon facilitated the launch of many more clinical trials. By 1986 enough data had been collected for the US Food and Drugs Administration to approve interferon for hairy-cell leukaemia. This was the first form of immunotherapy ever licensed for cancer.

5.6   ANTI-IDIOTYPE ANTIBODIES

Soon after Strander began testing interferon, another avenue opened up for immunotherapy, from the late 1970s. This was based on the work of Freda and George Stevenson, a husband and wife team of immunologists based at Southampton University in the UK. In 1975 they discovered that antibodies, a type of protein made by the immune system, could kill tumour cells by activating complement and other immune defence cells known as natural killer or NK cells. They proved this experimentally using serum raised in sheep immunised with cells taken from leukaemic guinea pigs.[36]

In addition to finding that the antibodies could mediate the destruction of tumours, the Stevensons found a unique marker 'protein' that could provide a very precise tool for targeting cancer. These proteins, known as idiotypes, appeared on the surface of malignant B lymphocytes (B cells) associated with leukaemia, lymphoma and multiple myeloma. The advantage of the idiotype was that it was only present on cancer cells and not normal cells.[36]

By 1980 the couple had successfully developed antibodies targeted at the idiotype for use in the first human subject, a 73-year-old man with chronic lymphocytic leukaemia (CLL). They had generated the anti-idiotype antibodies by injecting CLL cells taken from the patient into sheep. Encouragingly, the antibodies destroyed the man's leukemic cells by activating his complement. Despite this promising result, the Stevensons were limited in how far they could roll out the treatment to other patients because their process for manufacturing the antibodies was complex and time-consuming.[37]

More was achieved by Ron Levy, an oncologist at Stanford University. Inspired by the work of the Stevensons, he developed a way of producing anti-idiotype antibodies using a new technique developed by Cesar Milstein and Georges Köhler described in Chapter 4. Published in 1975, this method facilitated the large-scale production of standardised and highly specific antibodies called monoclonal antibodies (Mabs). By 1981 Levy had successfully treated his first patient with lymphoma. Subsequent patients treated with his anti-idiotype Mabs showed similarly positive results. Levy's treatment, however, had a major downside: the Mab had to be tailored to individual patients because it targeted surface antigens of the lymphoma tumours which were unique to each patient. Customising a Mab was inevitably time-consuming, taking up to six months. This was prohibitively expensive. Treatment of just one patient was estimated to cost $50 000. As Levy's sole source of financial support came from government and charitable bodies, progress was inevitably slow.[38–40]

5.7 INTERLEUKIN AND ADOPTIVE CELL THERAPY (ACT)

Soon after the first anti-idiotype antibodies were tested, news broke of the discovery of a new cytokine called interleukin 2 (IL-2). It was first detected by the American scientists Francis

Ruscetti, Doris Morgan and Robert Gallo in 1976.[41] IL-2 plays an important role in stimulating the growth of T lymphocytes (T cells), a type of white blood cell essential to the destruction of any foreign invader to the body.[42]

One of the first people to investigate IL-2 was Steven Rosenberg, the head of surgery at the US National Institute of Health. He was particularly interested in its potential to improve adoptive cell therapy (ACT). This was a technique that emerged in the 1960s following the observation that certain white blood cells, known as cytotoxic lymphocytes, destroyed cancer cells in test tubes. Initially, cancer patients were given white blood cells taken from volunteers who had been immunised with the patient's tumour cells. Such treatment, however, was severely hampered by the need for immunised donors and the difficulty of securing a subset of cytotoxic lymphocytes sufficiently powerful to attack tumours. The discovery of IL-2 radically changed the situation. Importantly, it provided a tool for growing the cells in a test tube.[43]

In December 1985, Rosenberg and his team published promising results in the *New England Journal of Medicine* for a new form of immunotherapy. It involved giving patients repeated infusions of a sub-population of cytotoxic lymphocytes directed against tumours isolated from peripheral human blood incubated with IL-2 in a test tube. These infusions were combined with several independent doses of IL-2. Given to 25 patients with advanced incurable cancer, the treatment reduced tumours by 50% or more in 11 of the patients. One patient with metastatic melanoma experienced a complete remission and remained free of cancer ten months after finishing the therapy.[44]

By the late 1980s Rosenberg had refined his therapy regime. His aim was to reduce the need for the use of IL-2 directly in patients as it had undesirable side effects. The new method involved taking lymphocytes obtained from the patient's tumours and incubating them with IL-2 (Figure 5.1). While shown to be promising in cancer patients, the process for cultivating the tumour infiltrating lymphocytes (TILs) was highly laborious and time-consuming as well as costly.[45] Over the following decades Rosenberg's and several other groups continued to experiment with the technique looking for ways to enhance the anti-tumour activity of the TILs. ACT, however, remained too complex and

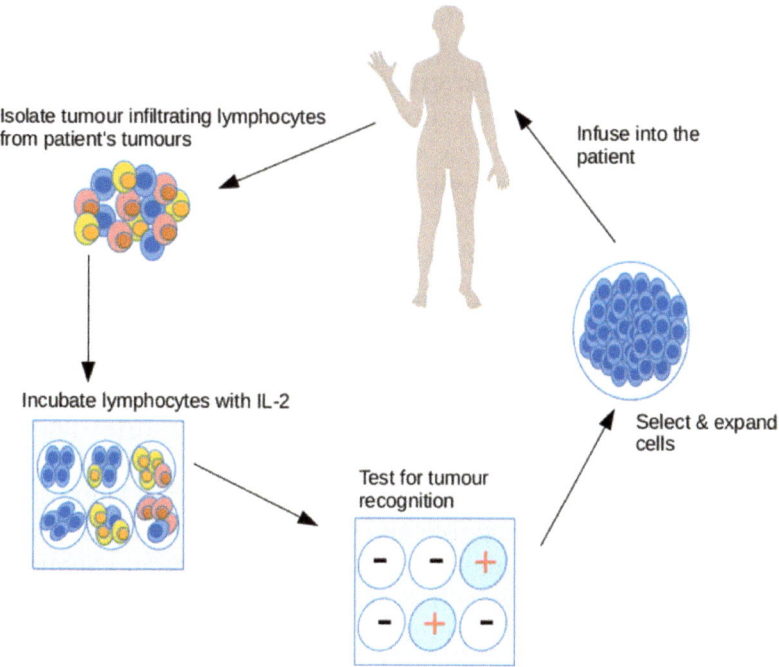

Figure 5.1 Adoptive cellular therapy for metastatic melanoma.

costly for its widespread adoption. A key drawback was that it required personalised treatment for each patient.[43]

While ACT gained little traction, recombinant IL-2 soon emerged as a treatment for advanced kidney and skin cancer. The cytokine was first approved in Europe in 1990 and the US in 1993. By this stage it was possible to produce genetically engineered forms of the cytokine. IL-2 is administered either as a single drug treatment or used in combination with chemotherapy or with other cytokines, such as interferon. The drug is used to enhance the efficacy of the other treatments. It can, however, cause intolerable side effects including fever and chills or flu-like symptoms, skin rashes, nausea and lowered blood pressure.

5.8 BCG VACCINE REVIVED

IL-2 was not the only form of cancer treatment approved in 1990. That year the FDA licensed the BCG vaccine for the treatment of superficial bladder cancer. This was a major comeback for the vaccine. As will be recalled, the reputation of the vaccine had been severely harmed by the Lübeck disaster of 1930. Nonetheless, it began to regain ground in the late 1940s as a public health tool for combating tuberculosis, a disease that resurged after World War II, and the development of mass vaccine campaigns against other infectious diseases.[46]

Use of the BCG as a cancer therapy was reignited in 1957 by the demonstration by Lloyd Old's team that the vaccine could activate macrophages and inhibit cancer in mice. By the 1970s researchers began to show that the BCG could be effective in combating cancer in humans. One of the first to demonstrate its potential was the haematologist George Mathé, at the Institut Gustave Roussy, Paris, who in 1969 published promising results from patients suffering from acute lymphoblastic leukaemia treated with a combination of BCG and bone transplants.[47] A year later an epidemiological study was published of children under the age of 15 in Quebec, Canada. This found that deaths from leukaemia between 1960 and 1963 were half as common among those vaccinated with the BCG than among those not vaccinated.[48] Another epidemiological survey of new-borns in Chicago between 1957 and 1969 similarly found that the vaccine helped protect against leukaemia and other forms of cancer. The vaccine was noted to reduce the mortality rate of cancer by as much as 74%.[49]

By the early 1970s a number of clinical trials were also indicating that the BCG vaccine could be an effective treatment for cancer. One study, carried out with 151 patients with malignant melanoma, was particularly encouraging.[50] The results, however, were inconclusive as the trials had not included any control groups. Subsequent controlled medical oncology trials showed the BCG provided no significant benefit. The disappointing results were attributed to the fact that most of those treated had been patients with advanced cancer who were expected to have a poor response to the vaccine.[51]

The discouraging results from the controlled trials greatly diminished the attractiveness of BCG for cancer therapy. Yet evidence soon emerged that the vaccine could help treat bladder cancer. Research in this area was led by Alvares Morales, a Canadian urologist. He devised the first protocol for its use in 1972. It involved administering the vaccine directly into the bladder through a catheter. The aim was to activate the body's immune cells to destroy the bladder's cancer cells. By 1976 Morales had effectively treated nine patients. Following this, the National Cancer Institute launched controlled human trials using Morales' technique. Initial results from the study, published in 1980, showed the BCG significantly reduced tumour recurrence in 54 patients and that its effect increased over time. By 1990 sufficient data had been collected from 2500 cases for the FDA to license the vaccine for the treatment of superficial bladder cancer. In 2010 the FDA further extended the use of the BCG by approving it for treating aggressive non-muscle non-invasive bladder cancer, one of the most prevalent forms of human cancer.[16]

5.9   A NEW CHAPTER FOR ACT

Soon after the FDA approved the BCG vaccine for aggressive bladder cancer, it approved another type of vaccine treatment for some men with prostrate cancer. This treatment, known as sipulecucel-T, uses dendritic cells, an important accessory of the immune system. The cells are harvested from the patient's blood and sent off to a central processing laboratory to be incubated with a type of protein made by combining a protein found on prostrate cancer cells with a growth factor. Following incubation, the cells are sent back to the clinic to be infused back into the patient. The aim of the treatment is to stimulate T cells to kill prostrate cancer cells.[52] Despite its approval, sipulecucel-T is unlikely to be used on a large-scale. This is, in part, because of its high cost, which reflects the complex process involved in the development of each treatment. One course of treatment costs $93 000 for each patient. The treatment also provides only a modest survival benefit.[53]

While sipulecucel-T has a number of drawbacks, its approval has ignited optimism for another type of ACT. This therapy builds on the original therapy protocol developed by Rosenberg in the late 1980s, which has been improved with the help of genetic engineering. The therapy, also examined in Chapter 10, involves extracting T-cells from a patient's blood and then genetically modifying them to express chimeric antigen receptors (CARs) on their surfaces. These are particular proteins that target specific antigens on tumour cells. Their presence improves the T cell's ability to bind to specific antigens and direct the body's own immune system to attack the tumours. Once made, the patient's genetically modified T cells are sent to specialised production facilities to be expanded and then returned to the hospital for infusion back into the patient. The first CAR T cells were developed in 1989 by Zelig Eshhar, an immunologist from the Weizmann Institute in Israel, while working with Rosenberg in his laboratory at the National Cancer Institute. They targeted human melanoma.[54]

Since then different academic and industrial groups have tweaked the basic design of CAR-T cells to increase their anti-cancer activity.[55] Clinical trials with the treatment are very encouraging, achieving a remission rate of up to 94% in patients who have not responded to other treatments. Some of the most promising CAR-T cell therapies are those that target CD19, a specific marker found on B cells related to blood cancers.[56]

As of March 2017 there were nearly 300 trials with CAR-T therapy, and two therapies were on track to be approved later in the year. Nonetheless, the treatment is still in its infancy and has potentially serious side effects, such as neurotoxicity and cytokine release syndrome. Furthermore, the process involved for developing the therapy is very expensive because it has to be personalised to each patient. Such therapy is predicted to cost between $300 000 and $500 000 per treatment. It would be one of the most expensive types of cancer treatments on the market.[57]

5.10   IMMUNE CHECKPOINT INHIBITORS

One of the most promising immunotherapies today is a class of drugs known as immune checkpoint inhibitors. These drugs target certain proteins, or receptors, found on the surface of T cells. Such receptors are known as checkpoint inhibitors because they help tame immune responses and prevent the destruction of healthy tissue. They enable the immune system to distinguish between normal 'self' cells and those perceived as 'foreign'. Other cells or molecules that bind to the receptors can either activate or switch off the action of the T cell. Some cancer cells find ways to bind to the receptors on the activated T cells to turn them off. This prevents them being attacked by the immune system.

One of the first inklings that cancer cells could block an immune response was noticed by two Jewish Austrian physicians, Ernest Freund and Gisa Kaminer, based at the Rudolf-Stiftung Hospital in Vienna.[58] In 1910 they reported that blood serum taken from healthy individuals could dissolve cancer cells, whereas that of cancer patients could not. They proposed that something in the cancer serum provided protection against the dissolving action of normal serum.[59] By 1924 they had found a substance in the intestines of cancer patients which when added to normal serum reduced its ability to dissolve cancer cells. Initially reported in the *Wiener Klinische Wochenschrift (Vienna Clinical Weekly)*, news of their finding was quickly picked up by *Time* magazine and publicised by newspapers worldwide.[60,61] Their discovery, however, was soon forgotten and in 1938, following the annexation of Austria by Nazi Germany, both physicians fled to London where they died soon after.

Many years later, in 1966, Karl and Ingegerd Hellström, a Swedish couple based at the Fred Hutchinson Cancer Center, began to investigate the reaction of lymphocytes to tumour antigens in test tubes. To their surprise, they observed that serum taken from mice with chemically induced tumours suppressed the reaction of lymphocytes. They attributed this to some sort of blocking factor.[62] In 1971 they published a paper in *Advances in Immunology* reviewing their work suggesting that 'blocking antibodies bind to the target tumour cells and thereby mask their antigens from detection by immune lymphocytes'.[63] By 1982 this paper had been cited 653 times, making it a citation classic.[64]

It would take some time before the exact blocking mechanism was unravelled. This was eventually pieced together as a result of a discovery made in 1987. That year a French group of researchers, led by Jean-Francoise Brunet, detected a new protein on the surface of T lymphocytes. They called the new molecule cytotoxic T lymphocyte-associated antigen 4 (CTLA-4).[65] For a number of years no one could work out what role CTLA-4 played in the immune system. The mystery was finally solved in 1996 by two teams working independently from each other: one led by James Allison at the University of California at Berkeley and the other by Jeffrey Bluestone at the University of California San Francisco. They showed that CTLA-4 could inhibit the activity of T cells.[66,67] Allison was the first to realise that the same mechanism could provide a means of treating cancer. To this end he developed a Mab to block CTLA-4. Encouragingly, his Mab inhibited the growth of tumours in mice. The University of California soon took out a patent on his technique.[68]

Soon after, Allison began to approach companies to see if they could help him develop his therapeutic concept further. Most, however, were reluctant to take on such a venture. In part this was because he was suggesting the suppression of a natural brake on the immune system to unleash an attack on the cancer. This contrasted with other forms of immunotherapy, most of which were designed to ramp up the immune system to attack cancer. Many immunology researchers were additionally sceptical that an antibody-based treatment could successfully treat solid tumours.[69]

The first people to run with Allison's proposition were Alan Korman and Nils Lonberg working at Medarex, a small bio-technology company originally set up in Princeton in 1987. By 1999 Medarex had secured the rights to the University of California's patent on the Allison technique. It had been first licensed to Nexstar Pharmaceuticals, where Korman had originally been based. Medarex was in a prime position to take the project forward because it already possessed mice with human antibody genes which could produce human Mabs. These mice had been genetically engineered through the use of a technology developed by Lonberg, and have been used for the development of numerous other treatments for cancer, inflammatory diseases and infectious diseases.[32,70]

By 2000 Medarex was ready to start its first clinical trials of a human Mab binding to CTLA-4. Until this point most cancer therapies had been judged according to how much they shrank a tumour. However, such criteria proved confusing for Medarex's drug. This was because some tumours appeared to increase in size. In the context of chemotherapy such tumour growth would have been taken as a sign that the treatment was not working. Further investigation revealed that some of the tumours only looked bigger because they were inflamed and full of invading immune cells, and the tumours eventually shrank months after treatment. New criteria thus had to be used for the evaluation of the new drug. This was helped by guidelines drawn up by the Cancer Vaccine Clinical Trial Working Group, an international consortium.[71] Based on both tumour shrinkage and an increase in the patients' overall survival,[72] in 2011 the FDA licensed Medarex's drug, ipilumumab, for the treatment of metastatic melanoma. It was the first immune checkpoint inhibitor to reach market.

Three years later, the FDA approved another immune checkpoint inhibitor, nivolumab, which had been developed by Medarex (which was acquired by Bristol-Myers Squibb in 2009). This was founded on the back of the discovery of another protein on T cells that could inhibit their activity. It was called PD-1 (programmed cell death protein 1). The protein had been first spotted in 1992 by Tasuku Honjo and his colleagues at Kyoto University.[73] However, its function remained an enigma until 2000, when Gordon Freeman and colleagues at the Dana-Farber Institute showed it to be an important mechanism for dampening the immune response after the elimination of a disease.[74] Two years later a Japanese team published a paper revealing that cancer cells were capable of hijacking the PD-1 protein to evade attack by the immune system.[75] By 2016 the FDA had approved three drugs that blocked PD-1 and another related protein PD-L1. In addition, more than a dozen trials had been completed in the US with such drugs for a range of cancers and more than 50 were underway.[76]

Several other immune checkpoint pathways are now being explored for the treatment of cancer (Figure 5.2). Work is now

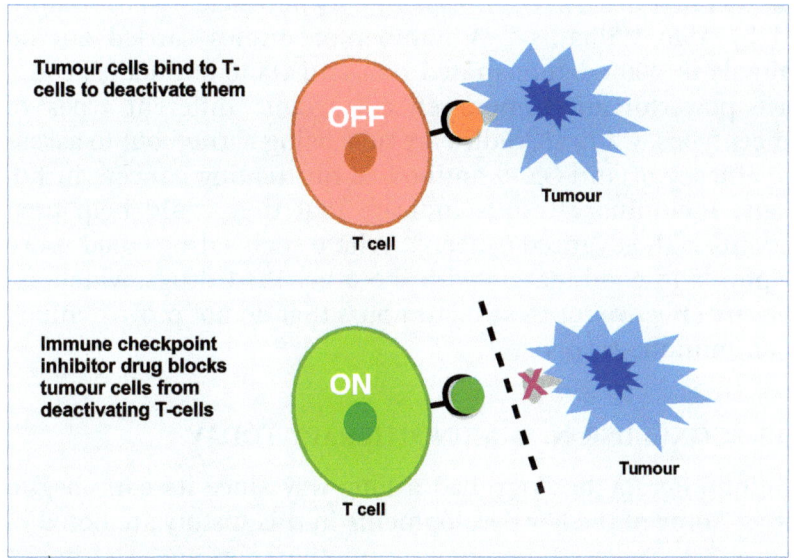

Figure 5.2 Basic diagram showing how an immune checkpoint inhibitor drug works.

underway, for example, to develop drugs that target the lymphocyte activation gene 3 (LAG3). This is a cell-surface molecule that has diverse biological effects on T cell function. It was first discovered by French researchers led by Frederic Triuebel at the Institut Gustave-Roussy in 1990.[77] Clinical trials with antibody-based drugs targeting LAG3 are already well underway with promising results.[78] Another immune checkpoint being investigated is T-cell immunoglobulin and mucin-domain containing-3 (TIM-3). This molecule was first identified in 2002 by Laurent Monney and his team at the Brigham and Women's Hospital in Boston and is now the subject of intense interest.[79,80]

5.11   STIMULATORY CHECKPOINT MOLECULES

As well as developing drugs that target immune checkpoints, work is now progressing on drugs that target stimulatory checkpoint molecules. The key molecules in this area are OX40, CD27, CD40, GITR and CD137. In contrast to the immune checkpoint inhibitors, which block negative signals to T cells, these newer drugs are designed to enhance T cell function. Some of the most advanced work in this area targets OX40. This protein was first discovered in 1987 by a team of scientists at Oxford led by Alan Williams.[81] A number of studies carried out on animals in 2000 demonstrated that anti-OX40 antibodies could have powerful anti-tumour effects against different types of cancer types. Clinical studies are now being carried out to assess the efficacy of anti-OX40 antibodies for treating cancers in humans. Preliminary results indicate that they could help treat patients with advanced cancer.[82] Where such a drug could prove helpful is in combination with the other PD-1 drugs, which are not very effective for treating tumours that do not provoke much of an immune response.[83]

5.12   CONCLUSION: IMMUNOTHERAPY TODAY

Immunotherapy has travelled a long way since its early beginnings. Some of the key developments in this history are noted in Figure 5.3. Today numerous immunotherapies are used in the clinic. Each uses different mechanisms to boost or restore the immune system's fight against cancer. Table 5.1 lists those

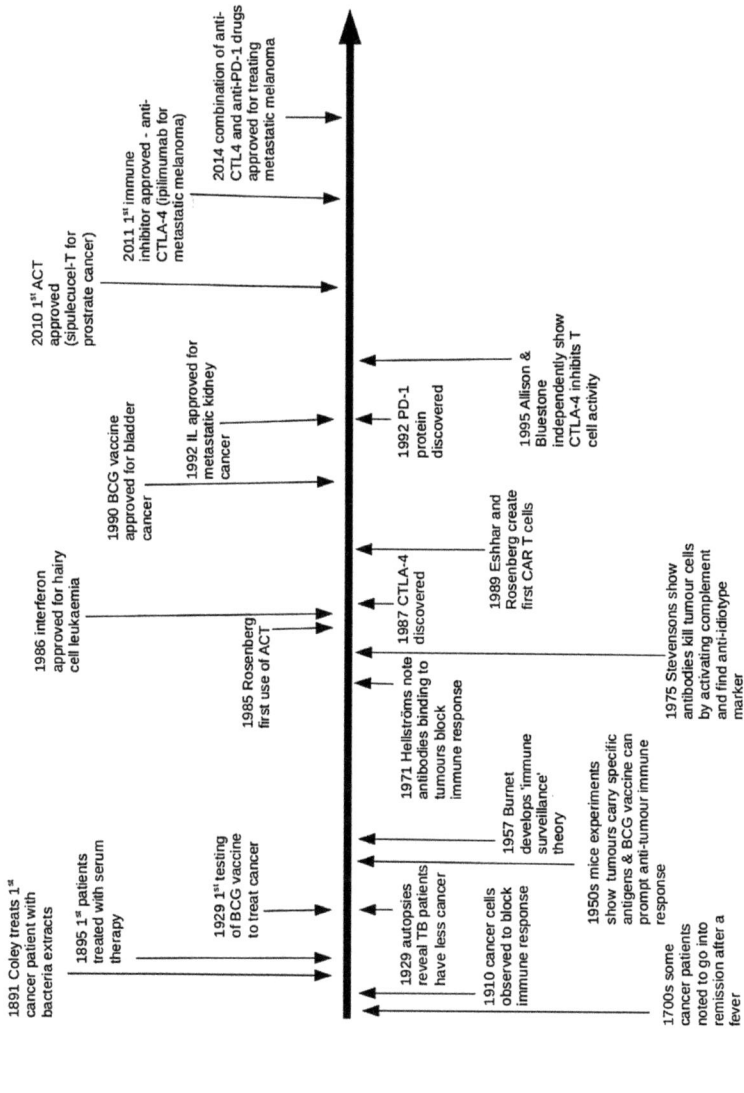

Figure 5.3 Timeline of some of the key developments in immunotherapy.

Table 5.1 Key FDA approved immunotherapies with advantages and disadvantages. Data taken from ref. 85 and 86.

Class	Type	Basic mechanism and major advantages	Major disadvantages	Approvals
Cytokines (proteins made by the immune system)	Interferon and interleukin are the two most common cytokines used in immunotherapy.	– Stimulates host's immune response	– Low response rates – Serious risk of systemic inflammation (Il-2) or high-dose toxicity (interferon)	1986, 1992, 1995, 1998
Vaccine	TB vaccine used to boost immune response to cancer cells	– Stimulates host's immune response – Minimal toxicity – Administered in outpatient clinic	– Lack of universal antigens – Ideal immunisation protocols lead to poor efficacy and response	1990
Mabs (laboratory-derived antibodies)	Used to bind to specific targets in the body to stimulate immune attack on cancer cells	– Binds very specifically to target – Used to treat broad range of cancers	– Expensive	1997, 1998, 2000, 2001, 2003, 2004. 2006, 2009
Immune checkpoint inhibitors (anti CTLA-4 and anti-PD-1 antibodies)	Blocks the ability of tumour cells to prevent an immune response. Often achieved with a Mab.	– Unleashes pre-existing anticancer T cell responses and possibly triggers new ones – Potent anti-tumour properties – Prolongs overall survival – Long-lasting clinical responses – Potential for broad range of cancers	– Only relatively small fraction of patients obtain clinical benefit – Severe immune-related adverse events observed in up to 35% of patients (anti-CTLA 4 drugs)	2011, 2014, 2015, 2016, 2017

Combination immunotherapy	Immune checkpoint blockade, key backbone	– Improvement of anti-tumour responses/Immunity	– May lead to increases in magnitude, frequency and onset of side effects	2016
Adoptive cellular therapy	Modification of patient's T cells to boost their ability to fight cancer	– Omits task of breaking tolerance to tumour antigens – High binding with effector T cells – Depleting lymphocytes prior to infusion enhances efficacy – Genetic T cell engineering could broaden treatment to malignancies other than melanoma	– So far limited to treating melanoma – Serious adverse effects – Lack of long-lasting responses in many patients – Requires time to develop desired cell populations – Expensive process to develop treatment because it is tailored to each patient	2010

approved by the FDA between 1986 and 2016. In addition to those approved, many are in the development pipeline. At the end of 2016 GBI Research analysts calculated that there were 2037 products in development, which equated to 37% of the total oncology pipeline.[84]

A large proportion of the treatments have been licensed in the last decade and contain Mabs. This includes the immune checkpoint inhibitor drugs. One of the reasons Mabs have proven so important to immunotherapy is because they are so versatile. Not only have Mabs proved helpful to blocking the mechanism tumour cells use to inhibit an immune response, but they can bind to tumour cells and thereby signal various immune cells to attack and destroy them. Similarly a Mab can be used to prevent a tumour cell getting access to growth factors or to inhibit the formation of new blood vessels that tumours need to grow.

The last 15 years has seen a major shift in the treatment options for cancer, aided by increasing knowledge of molecular and tumour biology. The success of immune checkpoint inhibitors in treating advanced metastatic cancer shows just how far our understanding of the multiple ways tumours manipulate the immune system to escape destruction has improved in recent years. Yet, advances in the field have not been straightforward and many have doubted its potential along the way. Moreover, immunotherapy still has far to go. For example, much more needs to be learned before maximum clinical benefit is achieved with the immunotherapies. One of the key research questions today is why some patients benefit from such treatment more than others. In addition, the therapies still carry the risk of serious adverse events. More than half of those who take immune checkpoint inhibitors, for example, experience severe side effects, some of which can be fatal.[87]

Cancer immunotherapy also faces an uphill struggle in dealing with a foe that is constantly mutating in response to therapy. Thus, while immunotherapy can induce complete and long-lasting cancer remission, some patients develop resistance to the treatment. Investigations have now been launched to see if such resistance can be resolved through the combination of different immune checkpoint inhibitors. This is already having promising results. For example, in 2015 it was reported that patients with advanced melanoma who were given a combination of the anti-PD-1 agent nivolumab with the anti-CTLA-4 agent ipilimumab had a median progression-free survival of 11.4 months, compared with 6.9 months if just given nivolumab and 2.9 months with just ipilimumab.[88,89]

Despite their promise, immunotherapy agents, like many other innovative drugs, are expensive. Drugs for cancer have traditionally tended to be more costly than those for other disease areas. Nonetheless, those with molecular targets, such as immunotherapies, are much costlier than those of the past. In 2015 the US National Bureau of Economic Research calculated that the overall price of cancer drugs had increased 10% every year between 1995 and 2013. In addition to this, the duration of treatment has increased. Some patients receive treatment for a number of years. On top of the costs of the drug there are the associated costs of diagnosis and nursing care. All of this is a significant burden on healthcare resources. For this reason major questions are now being raised about how to assess value of treatment versus survival benefit for patients and the long-term sustainability of the new cancer drugs.[90]

Cuba, a developing country strapped for resources, perhaps offers one way forward. It has found a means to produce cancer immunotherapies affordable for its whole population. Some idea of the headway it has made can be seen from the case of Cimavax, a vaccine treatment for non-small-cell lung cancer. The immunotherapy is the product of 25 years of collaborative research between Cuba's Center for Molecular Immunology (CIM) and Center for Genetic Engineering and Biotechnology. Cimavax is designed to stimulate the body's immune system to recognise and bind to a signalling protein known as epidermal growth factor (EGF) that spurs on the growth and proliferation of cells. Produced naturally in the body, some cancer cells encourage the body to make too much EGF so that the cells grow and divide uncontrollably. Cimavax aims to prevent EGF from attaching to receptors on cancer cells, thereby blocking the signal to grow and divide.[91] It has been shown to increase the survival of patients with advanced non-small-cell lung cancer in trials.[92,93] The drug received conditional approval in Cuba in 2009 and full approval in 2014. It is now distributed for free, like any other drug in the country. What is encouraging is that each shot of the vaccine costs the government just $1. Trials of the vaccine are now underway in Japan and a number of European countries. CIM is also testing it as a combination therapy with nivolumab in the USA in partnership with the Roswell Park Institute, New York.[94]

REFERENCES

1. J. Couzin-Frankel, Cancer Immunotherapy, *Science*, 2013, **342**(6165), 1432–1433.
2. A. Deidier, *Dissertation medicinal et Chirurgical sur les Tumeurs*, 1725.
3. W. Busch, *Klin. Wochenschr.*, 1868, **5**, 137.
4. F. Fehleisen, *Dtsch. Med. Wochenschr.*, 1882, **8**, 553–554.
5. P. Bruns, *Beitr. Klin. Chir.*, 1888, **3**, 443, cited in U. Hobhm, *Cancer Immunol. Immunother.*, 2001, **50**, 391–396.
6. W. B. Coley, *Am. J. Med. Sci.*, 1893, **105**, 487–511.
7. M. Tontonoz, The legacy of Bessie Dashiell, Cancer Research Institute, Dec 13 2013, http://www.cancerresearch.org/news-publications/our-blog/december-2013/the-legacy-of-bessie-dashiell. Accessed Jan 2017.
8. C. O. Starnes, *Nature*, 1992, **357**, 11–12.
9. I. Löwy, *Between Bench and Bedside*, Harvard University Press, 1996, p. 102.
10. R. Pearl, *Am. J. Hyg.*, 1929, **9**, 97–159.
11. G. W. Comstock, *Am. J. Epidemiol.*, 1999, **150**(12), 1263–1265.
12. I. Holmgren, *Schweiz. Med. Wochenschr.*, 1935, **65**(120), 1206.
13. I. Holmgren, *Acta Med. Scand.*, 1936, **90**(350), 3.
14. S. R. Rosenthal, *Am. Rev. Tuberc.*, 1936, **39**, 128.
15. N. M. Gandhi, A. Morales and D. L. Lamm, *BJU Int.*, 2013, **112**, 288–297.
16. H. W. Herr and A. Morales, *J. Urol.*, 2008, **179**, 53–56.
17. C. Bonah, *Soc. Hist. Med.*, 2002, **15**, 187–202.
18. H. Buchner, *Zentralbl. Bakteriol., Parasitenkd.*, 1891, **10**, 699–701.
19. E. Behring and S. Kitasato, *Dtsch. Med. Wochenschr.*, 1890, **16**(49), 113–114.
20. A. M. Silverstein, *Paul Ehrlich's Receptor Immunology: The Magnificent Obsession*, Academic Press, 2002.
21. A. B. Laurell, *Scand. J. Immunol.*, 1990, **32**(5), 429–432.

22. J. Hericourt and C. Richet, *C. R. Hebd. Seances Acad. Sci.*, 1895, **121**, 567–569.
23. M. Burnett, *Br. Med. J.*, 1957, 1(5023), 841–847.
24. C. Parish, *Immunol. Cell Biol.*, 2003, **81**, 106–113.
25. G. P. Dunn, A. T. Bruce, H. Ikeda, L. J. Old and R. D. Schreiber, *Nat. Immunol.*, 2002, **3**, 991–998.
26. R. T. Prehn and J. M. Main, *J. Natl. Cancer Inst.*, 1957, **18**, 769–778.
27. L. J. Old, D. A. Clark and B. Benacerraf, *Nature*, 1959, **184**, 291–292.
28. M. Burnett, *Br. Med. J.*, 1957, 1(5023), 779–786.
29. M. Burnett, *Br. Med. J.*, 1957, 1(5023), 841–847.
30. O. Stuman, *Adv. Cancer Res.*, 1975, **22**, 261–422.
31. A. Van Pel and T. Boon, *Proc. Natl. Acad. Sci. U. S. A.*, 1982, **79**, 4718–4722.
32. Email from N. Lonberg to L. Marks, 7 March 2017.
33. K. Sikora, *Br. Med. J.*, 1980, **282**, 855–858.
34. H. Strander, *Blut*, 1977, **35**, 277–288.
35. L. V. Marks, *The Lock and Key of Medicine: Monoclonal Antibodies and the Transformation of Healthcare*, Yale University Press, 2015, pp. 69–77.
36. G. T. Stevenson and F. K. Stevenson, *Nature*, 1975, **254**, 714–716.
37. M. Glennie and G. T. Stevenson, *Nature*, 1982, **295**, 712–714.
38. Anon, 2 Doctors win $50,000 each for work on a cancer drug, *The New York Times*, 4 Dec 1982.
39. M. Roth, Guided missile against tumors, *Pittsburgh Gazette*, 21 Oct 1985.
40. R. Levy and R. A. Miller, *Annu. Rev. Med.*, 1983, **34**, 107–116.
41. D. A. Morgan, F. W. Ruscetti and R. C. Gallo, *Science*, 1976, **193**, 1007–1008.
42. I. Löwy, *Between Bench and Bedside*, Harvard University Press, 1996, p. 129.
43. S. A. Rosenberg and N. P. Restifo, *Science*, 2015, **348**(6230), 62–68.
44. S. A. Rosenberg, M. T. Lotze and L. M. Muul, *N. Engl. J. Med.*, 1985, **313**, 1485–1492.
45. I. Löwy, *Between Bench and Bedside*, Harvard University Press, 1996, pp. 137–145.
46. S. Luca and T. Mihaescu, *Maedica*, 2013, **8**(1), 53–58.

47. G. Mathé, J. L. Amiel, L. Schwarzenberg, L. Schneider, M. Cattain, A. Schlumberger, J. R. Hayatt and F. De Vassal, *Lancet*, 1969, **I**, 697–699.

48. L. Davignon, P. Robillard, P. Lemonde and A. Frappier, *Lancet*, 1970, 638.

49. R. G. Crispen and S. R. Roisenthal, *Cancer Immunol. Immunother.*, 1976, **1**, 139–142.

50. D. L. Morton, F. R. Eilber, E. C. Holmes, J. S. Hunt, A. S. Ketcham, M. J. Silverstein and F. C. Sparks, *Ann. Surg.*, 1974, **180**(4), 635–641.

51. N. M. Gandhi, A. Morales and D. L. Lamm, *BJU Int.*, 2013, **112**, 288–297.

52. NCI, Cancer vaccines, https://www.cancer.gov/about-cancer/causes-prevention/vaccines-fact-sheet. Accessed Feb 2017.

53. K. Percia, J. C. Varela, M. Oelke and J. Schneck, *Rambam Maimonides Med. J.*, 2015, **6**(1), 1–9.

54. Z. Eshhar, *Hum. Gene Ther.*, 2014, **25**, 773–778.

55. V. Brower, The CAR-T Cell Race, *Scientist Magazine*, 1 April 2015, http://www.the-scientist.com/?articles.view/articleNo/42462/title/The-CAR-T-Cell-Race/. Accessed Mar 2017.

56. P. Hwu, CCR 20th Anniversary Commentary: Chimeric antigen receptors, *Clin. Cancer Res.*, 2015, 3099–3101.

57. A. Ward, Cancer therapy re-engineers cells to hunt and destroy, *Financial Times*, 2 June 2016.

58. R. W. Moss, History of checkpoint inhibition, 15 Feb 2017, http://www.ralphmossblog.com/2017/02/history-of-immune-checkpoint-inhibition.html.

59. E. Freund and G. Kaminer, *Biochem. Ztschr.*, 1910, **26**, 312; L. Herly, *Cancer Res.*, 1921, **6**(4), 337–356.

60. Anon, Chemistry of Cancer, *Time*, 3(4), 1 Jan 1928.

61. Anon, The Cancer scourge: Verge of discovery, *The Press*, 26 Feb 1924.

62. I. Hellström, K. E. Hellström, C. A. Evans, G. Heppaer, G. E. Piece and I. P. S. Yang, *Proc. Natl. Acad. Sci. U. S. A.*, 1969, **62**, 362–369.

63. H. O. Sjögren, I. Hellström, S. C. Bansal and K. E. Hellstöm, *Proc. Natl. Acad. Sci. U. S. A.*, 1971, **68**(6), 1372–1375.

64. K. E. and I. Hellström This week's citation classic, *Current Contents*, no. 20 (May 1982), garfield.library.upenn.edu/classics1982/A1982NN25500001.pdf.

65. J. F. Brunet, J. Denizot, M.-F. LLuciani, M. Roux-Dosseto, M. Suzan, M.-G. Matte and P. Golstein, *Nature*, 1987, **328**, 267–270.
66. T. L. Walnus, C. Y. Bakker and J. A. Bluestone, *J. Exp. Med.*, 1996, **183**, 2541–2550.
67. D. R. Leach, M. F. Krummel and J. P. Allison, *Science*, 1996, **271**(5256), 2734–2736.
68. J. P. Allison, *Cell*, 2015, **162**(6), 1202–1205.
69. Interview with Don Drakeman by Lara Marks, 10 March 2017.
70. U. Storz, *mAbs*, 2016, **8**(1), 10–26.
71. A. Hoos, G. Parmiani, K. Hege, M. Sznol, H. Loibner, A. Eggermont, W. Urba, K. Blumenstein, N. Sacks, U. Keiulholz and G. Nichol, *J. Immunother.*, 2007, **30**(1), 1–15.
72. J. D. Wolchchok, F. S. Hodi, J. S. Weber, J. P. Allison, W. J. Urba, C. Robert, S. J. O'Day, A. Hoos, R. Humphrey, D. M. Berman, N. Lonberg and A. J. Korman, *Ann. N. Y. Acad. Sci.*, 2013, **1291**(1), 1–13.
73. Y. Ishida, Y. Agata and T. Honjo, *EMBO J.*, 1992, **11**, 3887–3895.
74. G. J. Freeman, A. J. Long, Y. Iwai, K. Bourque, T. Chernova, H. Nishimura, L. J. Fitz, N. Malenkovich, T. Okazaki, M. C. Byrne, H. F. Horton, L. Fouser, L. Carter, V. Ling, M. R. Bowman, B. M. Carreno, M. Collins, C. R. Wood and T. Honjo, *J. Exp. Med.*, 2000, **192**(7), 1027–1034.
75. Y. Iwai, M. Ishida, Y. Tanaka, T. Okazaki and T. Honjo, *Proc. Natl. Acad. Sci. U. S. A.*, 2002, **99**(9), 2293–2297.
76. L. Anderson, Immune checkpoint inhibitors: Boosting the cancer battle, 22 Nov 2016, https://www.drugs.com/slide show/immune-checkpoint-inhibitors-1249. Accessed Feb 2017.
77. F. Triebel, S. Jitsukawa, E. Baixeras, G. Roman-Roman, C. Genevee, E. Viegas-Pequinot and T. Hecend, *J. Exp. Med.*, 1990, **171**, 1393–1405.
78. P. Sharma and J. P. Allison, *Cell*, 2015, **161**, 205–214.
79. L. Mooney, C. A. Sabatos, J. L. Gaglia, A. Ryu, H. Waldner, T. Chernova, S. Manning, E. A. Greenfield, A. J. Coyle, R. A. Sobel, G. J. Freeman and V. K. Kuchroo, *Nature*, 2002, **415**(6871), 536–541.
80. A. C. Anderson, *Cancer Immunol. Res.*, 2014, **2**(5), 393–398.

81. D. J. Paterson, W. A. Jeffries, J. R. Green, M. R. Brandon, M. P. Corthesy, M. Puklavec and A. F. Williams, *Mol. Immunol.*, 1987, **24**(2), 1281–1290.

82. B. D. Curti, M. Kovacsovics-Bankowski, N. Morris, E. Walker, L. Chisholm, K. Floyd, J. Walker, I. Gonzalez, T. Meeuwsen, B. A. Fox, T. Moudgil, W. Miller, D. Haley, T. Coffey, B. Fisher, L. Delanty-Miller, N. Rymarchyk, T. Kelly, T. Crocenzi, E. Bernstein, R. Sanborn, W. J. Urba and A. D. Weinberg, *Cancer Res.*, 2013, **73**(24), 7189–7198.

83. J. de Lartigue, OX40 agaonists forge a path in combination immunotherapy, *OncLive*, 2 March 2017, **18**(5), http://www.onclive.com/publications/oncology-live/2017/vol-18-no-5/ox40-agonists-forge-a-path-in-combination-immunotherapy?p=3. Accessed Mar 2017.

84. *Clinical Leader*, 14 Dec 2016, https://www.clinicalleader.com/doc/cancer-immunotherapies-space-by-says-gbi-research-0001. Accessed Mar 2017.

85. Fight Cancer with Immunotherapy, State of cancer immunotherapy, http://www.fightcancerwithimmunotherapy.com/immunotherapyandcancer/stateofcancerimmunotherapy.

86. S. Farkona, E. P. Diamandis and I. V. Blaustig, *BMC Med.*, 2016, **14**, 73.

87. F. Kroschinsky, F. Stölzel, S. von Bonin, G. Beutel, M. Kochanek, M. Kiehl and P. Schellongowski, *Crit. Care*, 2017, **21**, 89.

88. N. P. Restifo, M. J. Smyth and A. Snyder, *Nat. Rev.: Cancer*, 2016, **16**, 121–126.

89. A. Andrews, *Am. Health Drug Benefits*, 2015, **8**, 9.

90. Cancer Research UK, Health economics: The cancer drugs cost conundrum, 10 Aug 2016, http://www.cancerresearchuk.org/funding-for-researchers/research-features/2016-08-10-health-economics-the-cancer-drugs-cost-conundrum. Accessed Mar 2016.

91. Cancer Research UK, Can you tell me about CimaVax lung cancer vaccine? http://www.cancerresearchuk.org/about-cancer/cancers-in-general/cancer-questions/can-you-tell-me-about-the-cimavax-lung-cancer-vaccine. Accessed Apr 2017.

92. E. Neninger Vinageras, A. de la Torre, M. Osorio Rodríguez, M. Ferrer Catalá, I. Bravo, M. Mendoza del Pino, D. Abreu Abreu, S. Acosta Brooks, R. Rives, C. del Castillo Carrillo,

M. González Dueñas, C. Viada, B. García Verdecia, T. Crombet Ramos, G. González Marinello and A. Lage Dávila, *J. Clin. Oncol.*, 2008, **26**(9), 1452–1458.

93. P. C. Rodriguez, X. Popa, O. Martínez, S. Mendoza, E. Santiesteban, T. Crespo, R. M. Amador, R. Fleytas, S. C. Acosta, Y. Otero, G. N. Romero, A. de la Torre, M. Cala, L. Arzuaga, L. Vello, D. Reyes, N. Futiel, T. Sabates, M. Catala, Y. Flores, B. Garcia, C. Viada, P. Lorenzo-Luaces, M. A. Marrero, L. Alonso, J. Parra, N. Aguilera, Y. Pomares, P. Sierra, G. Rodríguez, Z. Mazorra, A. Lage, T. Crombet and E. Neninger, *Clin. Cancer Res.*, 2016, **22**(15), 3782–3790.

94. E. Schumaker and A. Almendraia, Cuba's had a lung cancer vaccine for years, and now it's coming to the U.S., *Huffington Post*, 15 April 2016, http://www.huffingtonpost.com/2016/02/22/cuba-lung-cancer-vaccine_n_7267518.html. Accessed Apr 2017.

CHAPTER 6

Gene Therapy: An Evolving Story

COURTNEY ADDISON

University of Copenhagen, Denmark
Email: cpa@ifro.ku.dk

6.1 INTRODUCTION

In July 2015, I found a ten-year old East African girl sitting on a sixth floor bed in a British children's hospital. She wore a fluffy purple cardigan, a bow around her head, and her father was keeping an eye on her from across the room. Isla has a rare disease called adenosine deaminase severe combined immune deficiency (ADA-SCID), which prevents her immune system from working properly. The disease is caused by a problem with the protein-coding gene known as adenosine deaminase (ADA). It is an autosomal recessive condition, meaning that both of Isla's parents had to have one copy of the defective gene; when two parents are carriers for such diseases, their chances of having an affected child, like Isla, are one in four. The first course of treatment for children with ADA-SCID is a bone marrow transplant, but that relies on finding a good donor match; none had turned up for Isla so she was in hospital to receive gene therapy. When I left the hospital Isla was well, but it would take another year before we would know the full outcome of her treatment.

Engineering Health: How Biotechnology Changed Medicine
Edited by Lara V. Marks
© The Royal Society of Chemistry 2018
Published by the Royal Society of Chemistry, www.rsc.org

Shortly after, in September 2015, the social media was buzzing with the announcement that Elizabeth Parrish, a charismatic forty-four year old who heads up an American biotechnology start-up company, had become the first person to receive another form of gene therapy - to reverse aging. Parrish was given two injections. One contained the gene follistatin, known to increase muscle mass in animals. The other provided genes to help produce telomerase, a protein that extends telomeres, a component of chromosomes known as the 'aging clock'. Experiments by other research groups have shown that injecting telomerase could extend the lifespan of mice. News of Parrish's treatment sparked major discussion within the scientific and medical community. While some equated it with medical quackery, others believed it heralded a new era in gene therapy, which would no longer be confined to treating a disease but would also be used to reverse normal processes like aging. What added fuel to the debate was the fact that Parrish had received her treatment in Columbia so as to bypass US regulatory authorities.

While the two examples above seem worlds apart from each other they help demonstrate the range of activity that falls under the banner of 'gene therapy', and the limitations of discussing 'gene therapy' as a single entity. In reality, the term 'gene therapy' is contested and used in widely divergent ways. Various labels have been deployed to describe this process, including gene therapy, gene transfer, genetic medicine, gene editing and gene surgery. At its most basic level the term refers to a set of strategies to modify an individual's genes for medical purposes.

Numerous techniques can be described as gene therapy. It ranges from giving a whole gene to a patient, known as gene addition, with the aim of curing their disease, to exon skipping, which encourages cells to skip sections of genetic code so as to create a partially functional protein despite the presence of a genetic mutation. Unlike gene addition therapy, exon skipping does not make a permanent change to the patients' genome. Another technique involves genetically engineering T cells, important players in the immune system, so that they can better recognise and kill cancer cells. Known as CAR-T cell therapy, explored in chapter 5, this method is looking to be a promising treatment for leukaemia.

In this chapter I concentrate on gene therapy in the form of whole gene addition used to medicate disease. The mechanics of how this works are introduced in the next section. The aim here is, first, to give the reader a grounding in how gene therapy works in practice, and, second, to show how varied gene therapy can be, according to the disease being treated and the body part targeted. As the chapter proceeds, I will trace how a whole scientific field came to cohere around a particular set of molecular techniques, which quickly diversified into a range of disease-specific gene therapies practiced today. This process, as I show, has in itself generated specific ethical debates, with various parties asking what types of gene therapy are acceptable and on what grounds. Following this I provide an overview of the present state of this field, highlighting new advances in gene editing and developments in mitochondrial donation.

By the end of the chapter I hope readers will have a sense of how the development of gene therapy has been a process of negotiating complex social and ethical as well as scientific issues.

6.2 HOW GENE THERAPY WORKS

The premise of gene therapy is simple. If certain diseases are caused by genetic mutations, then removing or overcoming the mutation should cure the disease. Adding a healthy version of the malfunctioning gene seems like an obvious and elegant way to achieve this. However, as with most medical matters, the reality is much more complex than it appears at first. This section begins with a basic introduction to genes and genetic disease. Readers who are familiar with this material may wish to skip ahead. From this foundation, I explain how gene therapy is supposed to work, and then how it actually works—or not.

6.2.1 How Genes Cause Disease

One cell of the human body (say from your skin, hair or internal organs) contains all of an individual's genetic material. Most of this is found in the cell's nucleus, and it comes wound up in a helix resembling a twisted ladder, each step of which contains a pair of complementary chemical sub-units, known as nucleotides or bases: adenine, thymine, guanine and cytosine. The first two always go together, as do the last two, making base pairs of A–T and C–G. Over three billion base pairs are stacked into this ladder for every person, and the string of letters they make up is that person's genome. Some strings of letters are genes, each of which can be millions of base pairs long. There are some 20 000 genes in the average persons' genome, interspersed with regions that are said to be 'non-coding'. Each gene codes for a protein, which is achieved through a process called transcription. During transcription an enzyme, known as RNA polymerase, works its way along a segment of DNA and copies out the list of base pairs to make a strand of almost identical messenger RNA (mRNA). It reads the DNA in chunks of three base pairs at a time—each base pair triplet is called a codon and codes for an amino acid. The mRNA then attaches itself to a ribosome, a complex of molecules, which reads the chain of bases in chunks of three and attaches the matching amino acids. By the time the ribosome reaches the end of the gene segment, a string of amino acids has been assembled, and these fold up precisely into a knot. That knot is a protein. Each protein is made up of lots of amino acids.

The above is a very rudimentary description of how the human genome codes for proteins, which perform essential life-sustaining functions in the body. It provides a sufficient foundation, however, for understanding the relationship between genes and proteins and their role in disease and the fundamental principles behind gene therapy, namely the role of genes and proteins in disease. Mutations in the genome can create a chain reaction: the mRNA that is crafted during transcription may reflect an error in the original DNA strand; the ribosome will be given a wrong codon and may attach the wrong amino acid; this in turn may lead to a protein misfolding and functioning incorrectly, or not at all.

Cystic fibrosis (CF) is a good example of a genetically caused disease. It is a systemic disease, meaning it affects many parts of the body, but it is best known for the thick mucus that builds up in patients' lungs and airways, obstructing their breathing. The disease originates from the CFTR gene, which is made up of nearly 200 000 base pairs. Within this gene, 2000 known mutations can cause CF, though most patients share the F508 delta deletion. This means that one of those almost 200 000 base pairs is absent, causing severe and eventually lethal disease. Gene therapy for CF aims to give patients a new, healthy copy of the CFTR gene. The appeal of gene therapy for this condition is that, in theory, it ought to work regardless of which mutation the CFTR gene has, meaning that any patient could benefit from the treatment. Because a whole new gene is being introduced to the body, the action (or inaction) of that entire mutated gene can be fixed. In practice CF gene therapy has been extremely difficult, for reasons I will explore shortly.

6.2.2 How Genes (Might) Fix Disease

To date, the most common type of gene therapy that has been attempted has been gene addition therapy. Like that described for CF, this means giving a whole gene to a patient to compensate for the malfunction of their existing ones. A clone of the desired gene can either be made or bought; many human and animal genes are available for purchase from 'libraries' housing thousands of DNA segments. Once obtained, modifications can be made to the gene to make it more effective when administered. For example, many researchers add a specific region of DNA, known as a promoter region, to prompt the cell to begin transcribing the gene. They may also pursue a strategy known as codon optimisation. Certain codons (the three base pair chunks that code for amino acids) are more rapidly and accurately translated by certain cells than others. It is possible to exchange some slower codons for faster ones without affecting the amino acid produced, making them better suited to the cell they will be delivered into. These and other changes can be engineered in the laboratory to make a given gene therapy more effective once administered.

The actual administration of gene therapy poses a significant challenge. Getting a therapeutic gene into a patient is most commonly achieved by using a vector, a molecule that binds to the DNA and helps carry it into the body. By 2015 approximately 66% of vectors used in gene therapy clinical trials were viral.[1] Viruses are used because they have evolved to get into human cells. Certain parts of the genomes of the viruses are removed when used in gene therapy so that they will not damage the host.

A range of viruses have so far been used in the field. Each type of vector has different advantages and drawbacks, and they are selected for a particular task accordingly. Factors to consider include the carrying capacity of the vector (*i.e.* how large a gene can it hold), the tissue type being targeted (will the vector go to dividing or non-dividing cells) and the potential genotoxicity of the vector (whether it will integrate into parts of the genome where it might turn on, for example, a tumour-producing gene).

Lentiviruses, derived from the HIV-1 virus are one of the most popular vectors used. The advantage of such viruses is that they can carry a relatively large genetic load (up to 10 kilobases), and tend to integrate in relatively safe regions of the genome. Because the wild type virus causes HIV infection, a section of the virus' genome is routinely removed to make it self-inactivating, and they are widely regarded as an acceptably safe choice.[2] Although wild-type lentiviruses will only target certain types of tissue, an 'envelope' can be added to make them bind to a wider range of tissues. Lentiviruses have been used with some success for gene therapies targeting lethal childhood conditions including adrenoleukodystrophy[3] and ADA-SCID.[4]

Non-viral delivery strategies are less commonly used, despite the fact that virus-free approaches are much less likely to antagonize the immune system. The limited use of non-viral methods is largely due to the reduced efficiency of these approaches as compared with viral vectors; put simply, non-viral strategies tend to deliver smaller amounts of the therapeutic gene, and that which is delivered remains for less time.[5] Several methods of non-viral gene delivery are available. The desired DNA segment can be administered 'naked', or without an encompassing molecule. Lipids (*i.e.* fat) can also be used to carry the gene into the body.

6.2.3 Technical Challenges for Gene Therapy

Once the therapeutic gene/s and vector have been assembled, they need to be delivered to the patient. The route taken for the administration of the gene therapy varies considerably. Among the issues to be confronted is what part of the body will be the most effective site for the gene therapy to propagate itself from. Which site is chosen largely depends on the disease being treated.

The difficulties involved are illustrated by the case of CF. Early on gene therapy was deliberately delivered to patients' lungs and airways *via* inhalation. This was because CF was initially thought to be primarily a disease of the lungs, and the lungs, in turn, were envisioned as quite an accessible organ (we all breathe into them, after all). However, patients' airways were blocked by sticky mucus. This made it hard for the gene therapy to penetrate. The size of the lungs also meant a very large quantity of therapeutic was needed to have an effect. Added to this was the fact that CF is not only a lung disease—it is systemic. This meant that treating all of the symptoms required a therapy that gets beyond the lungs.[6]

Alternative administration strategies have been used to treat other diseases. In the case of severe immune deficiencies, for example, (*e.g.* ADA-SCID, Wiskott–Aldrich Syndrome) patients' stem cells have been extracted from their bone marrow, and genetically modified in the laboratory before being returned to the patient in much the same way as a standard bone marrow transplant. Once back in the body these cells replicate, each one splitting into two, with some of the daughter cells carrying the therapy. Another strategy is that used for eye diseases. Some retinal gene therapies involve giving an injection directly into the eye, where the blood–ocular barrier prevents it getting into the rest of the body. One of the advantages with the eye is that there are two of them. This allows the untreated eye to act as a control to compare the treatment against.

Negotiating the body is often a steep learning curve, and it can take some trial and error to determine how different bodily organs and tissues respond to gene therapy. This is but one of the challenges that gene therapists have faced in developing treatment. Contending with the immune system is another major issue. Where this manifests itself is in the context of the viral vector used to deliver the therapy. Just as viruses have evolved to get into bodies, so our bodies have evolved an immune response to keep them out. Removing certain sections of the viruses' genomes has, in some cases, helped to reduce their ability to provoke an immune response. Another problem is that retroviruses can randomly integrate into the chromosome. Disturbingly, this was seen to cause leukaemia in several patients in the early 2000s. Since then, it has become routine to remove certain genes from such viruses and the range of viruses used as vectors has diversified. This has diminished the risk of accidentally activating cancer, although the problem has not entirely disappeared.

Perhaps the most significant challenge has been working out how to regulate the gene once delivered, to ensure it has long lasting effects (expression). This problem has been negotiated by various means, including engineering the DNA in ways that enhance the therapeutic gene's efficacy (*e.g.* adding promoter regions, codon optimizing, *etc.*), and improving the choice of vectors for particular cell types.

Of course, in addition to all of these technical matters, researchers have to find gene therapy solutions that are acceptable to their patients. Something as specific as a lentivirus can be overwhelmed by the spectre of HIV. How important this can be was made clear to me by a man I interviewed who suffered from haemophilia, an inherited condition that reduces the body's ability to clot. He reminded me that during the 1980s some 50% of haemophilia patients were infected with HIV after receiving blood transfusions.[7] Though an ardent supporter of gene therapy, the memory of this terrible event left him unable to bear the thought of lentiviral therapy, regardless of whether the infectious components of the virus were removed or not.

6.3 THE EMERGENCE OF THE GENE THERAPY FIELD

When and how did gene therapy begin? The first legal application of gene therapy in humans took place in 1990. However, its origins are rooted in work done in the nineteenth century by the likes of Charles Darwin and Gregor Mendel, to understand the processes of evolution and genetic inheritance. The rise of the eugenic movement in the early twentieth century was also a formative influence. Embraced by a number of physicians and geneticists, eugenics aimed not only to improve the genetic quality of the human population through selective breeding, but also to alleviate unnecessary suffering.[8] Ideas about genetics and disease entered a new phase with James Watson and Francis Crick's determination of the double helix structure of DNA in 1953. By the early 1960s the terms 'gene therapy' and 'gene surgery' were being used—perhaps, one commentator suggests, as a way of distinguishing gene therapy from eugenics.[9] The eugenic legacy was something of a spectre in the early days of gene therapy, but was kept at bay through a series of ethical debates and a resolute fixation on disease as the appropriate target of gene therapy. In 1967 Marshall Nirenberg, a Nobel Prize-winning geneticist, stated that cells would be genetically programmable within 25 years. At the same time he warned that the consequences of such technology should be considered before what could be done outpaced what ought to be done, or avoided. "When man becomes capable of instructing his own cells, he must refrain from doing so until he has sufficient wisdom to use this knowledge for the benefit of mankind," he stated.[10]

6.3.1   The Scientific Groundwork is Laid

By the late 1960s several important laboratory techniques had begun to be brought together that helped lay a foundation for gene therapy. The first was the ability to create custom cell lines. This was aided by a new ability to keep cells alive in culture indefinitely, developed in the first half of the 20th century. Prior to this bits of tissue could survive only a few weeks once outside the body.[11] A second crucial development was the identification of site-specific restriction enzymes, small molecules that can cut DNA at specific locations, in the late 1960s and early 1970s. These enzymes provided the means to not only cut DNA, but to recombine it in new ways. Researchers proceeded to do so with vigour, combining DNA segments from different species and inserting them into bacteria or their immortal cell lines to see how they behaved. Paul Berg and two colleagues conducted the first such experiment, joining viral DNA from bacteria with viral DNA from a primate, and inserting this hybrid into viral and bacterial cells.[12]

6.3.2   Ethical Questions Are Raised

The ability to muddle the boundaries between species and in particular to create genetically modified viruses, prompted concern among scientists and the public alike. In 1974 eleven scientists proposed a moratorium on certain kinds of research with recombinant DNA (rDNA), also known as genetic engineering, and in 1976 this same group held a meeting of experts at Asilomar in California. There they decided that rDNA research should go ahead, but that there should be strict regulations in terms of containing modified organisms and ensuring good laboratory practice.[13,14] The recommendations of the Asilomar group were to be enforced by the establishment of a rDNA Advisory Committee (RAC) by the National Institutes of Health (NIH).

Following the Asilomar meeting researchers returned to their work, and began experimenting with genetic engineering in a wider range of cells (including from yeast, plants and animals), bacteria and, in some cases, complete non-human organisms.[12] Prior to this point most of this work had been government funded, but as the 1970s drew to a close commercial interest in rDNA technology began to rise and it began to be deployed for the mass production of previously scare natural therapeutic proteins such as insulin.

The first application of gene transfer in humans occurred in 1980. This was carried out by the American clinician Martin Cline on patients in Italy and Israel suffering from beta-thalassemia, a blood disorder that usually results in a young death for those affected. Known already to be caused by a defect in the beta globin gene, Cline attempted to extract the patients' bone marrow and insert a beta globin gene bound to another that would give it a selective advantage once in the cells, both packaged into a herpes virus vector.[15] Cline's action sparked immediate concern as he had proceeded without waiting for approval to perform the same technique at UCLA, his home institution. Moreover, earlier animal models offered little indication that his procedure would be successful.

Cline's experiment not only failed in the two patients, but he was thoroughly chastised by colleagues and regulators. The director of the National Institutes of Health (NIH), who commissioned an investigation into the case, stated that, "Dr Cline has violated both the letter and the spirit of proper safeguards to biomedical research".[16] From then on the NIH made it non-negotiable to get approval from the RAC before proceeding with any gene transfer experiments. The public denouncement of Cline marked an important moment for the young gene therapy community. Cline was paraded as an example of what regulators considered right and wrong in the field and to stress the need to secure peer review of research plans before they were implemented. The incident, more than any other, opened up the question of how gene therapy might be ethical and socially acceptable.[17]

As Cline's mistakes were unfolding in the public eye, the US government received a letter penned jointly by Catholic, Protestant and Jewish leaders.[18] They were concerned that changing human genes was a more ethically fundamental question than was being acknowledged, and also questioned how such work could be controlled effectively and marketed responsibly. The trio called on President Carter to begin a public conversation on the matter, which resulted in his appointing a commission to further investigate. Two years later that Commission produced a report called 'Splicing Life', which concluded that the science behind gene transfer was acceptably sound, being little different from the natural splicing of genes that bacteria carry out. The commissioners concluded the technology could be put to worthwhile use with dedicated oversight.[19]

Ultimately the Commission felt the potential to cure disease and alleviate suffering was sufficiently compelling to continue genetic engineering research. However, they cautioned that, "gene splicing could have far-reaching consequences that could alter basic individual and social values".[19] Interestingly, they raised ethical concerns overlooked by scientists in earlier deliberations on genetic engineering. Among the questions they asked were: Should we meddle with genes for reasons other than disease? Would gene transfer alter fundamental characteristics of human nature? Who gets to decide how this progresses, and who is entitled to benefit from it? Should private companies be able to profit from taxpayer-funded basic research? Such questions would become abiding queries of bioethicists for years to come.

Some of the questions raised in 'Splicing Life' found resolution almost immediately, most notably at a hearing convened by Al Gore, then a US congressman. Gore assembled a group of scientists, scholars and legal experts to discuss a wide range of questions about genetic engineering. A contributor to this event was W. French Anderson, a key scientist working with gene transfer who had been considering the prospects of gene therapy for some time. As early as 1972 Anderson favoured gene therapy as a means of solving diseases that originated in theoretically replaceable errors in the DNA code for which there was no treatment. However, he was concerned about where to draw the line for treatment. As he put it, "Suppose the technique proves itself to be totally safe and reliable. Suppose it becomes possible to insert any gene or genes one might want into human cells, including germ (reproduction) cells. Then who gets to decide what are 'good' uses and what are 'bad' uses?"[20]

Anderson proposed that there were in fact two dividing lines. One ran between gene transfer for disease and gene transfer for enhancement. He argued that using gene therapy to treat seriously sick patients was justifiable, whereas using the technology to adjust superficial features such as height, intelligence or hair colour was not. The second line ran between gene therapy carried out on somatic and germ line cells. Somatic cells are body, or tissue, cells that make up skin, organs, muscle, blood *etc.* Genetic changes made to such cells cannot be passed on to subsequent generations. Germ line cells are the reproductive cells, sperm and egg cells, which after fertilization become a foetus. In contrast to somatic cells, any genetic changes made to an individuals' germ line cells are likely to be passed on their offspring. Anderson argued that germ line gene therapy was unacceptable, but might become feasible in later years to treat severe inherited diseases.[21] He pointed out, however, that germ line therapy would involve making changes to a person who could not offer their consent (being unborn), so posed further ethical questions. The two lines suggested by Anderson quickly became convention in gene therapy and have been disturbed only very recently.[22]

6.3.3 Preparations Begin for First-in-human Gene Therapy

Pre-clinical gene therapy experiments increased during the 1980s. Richard Mulligan, a scientist working at MIT, developed the first retroviral vector, which was quickly adopted by the gene therapy community.[23] Mulligan followed this achievement by combining his retrovirus with a piece DNA taken from *Escherichia coli* bacteria, which worked as a marker and when administered to a cell population allowed him to identify how many cells had picked up the gene transfer. During this same period Anderson himself tried to gather the necessary resources to treat patients with a rare immune condition known as ADA-SCID. He acquired a clone of the ADA gene from a friend, and enlisted the help of a paediatrician, Michael Blaese, who had the necessary disease expertise. Anderson then founded Genetic Therapy Inc., one of the field's first companies, with investment from a venture capitalist.[23]

The effort of earlier decades culminated in 1989 when Anderson, Blaese, and colleagues received NIH approval to perform the first gene transfer in human subjects.[24] The experiment did not constitute 'gene therapy' in the sense of transferring a corrective gene for treatment. Rather its purpose was to test how much of a gene could be transferred to a cell and for how long, as well as the side-effects of the retroviral vector. The experiment involved the administration of tumour infiltrating lymphocytes (a type of blood cell that can get into a solid tumour) that had been genetically modified to express a marker gene to five people with advanced cancer.

Not long after, the same team was authorized to conduct the first gene therapy trial. In 1990 they treated two children suffering from ADA-SCID with a healthy version of the ADA gene, packed in a retroviral vector.[25] The children responded well, showing no dangerous response to the vector. Unsure of how the procedure would go, the team had chosen to keep the children on Peg-ADA, an enzyme replacement therapy that can have restorative effects for patients. For this reason it was hard to tell if the improvement in the children's condition was due to the gene therapy or the Peg-ADA.

Successful or not, approval of the trial eased the way for others. By the end of 1991 the NIH had given the green light to several other projects by other teams, and trials began in Europe not long after.[26] The ADA-SCID trial also ignited a new interest in the field among previously wary biotech and, notably, larger pharmaceutical companies. In 1991, for example, Sandoz Pharma, Ltd. bought $10 million of Genetic Therapy Inc.'s stocks, giving the small company financial assets of over $20 million by the year's end.[27]

6.3.4 Gene Therapy Grows Up

Gene therapy activity grew throughout the 1990s and diversified to address a range of medical problems. This was accompanied by a lot of hype in the media, which frequently underplayed the complexity and uncertainty of gene therapy. More private companies became involved in the field over the ensuing decade. Such companies increasingly guided work towards conditions that had more potential to offer returns. The early focus on rare diseases, often caused by single genes, soon moved towards a new focus on cancers. By 1996 70% of gene therapy trials were for cancer.[28] The interest of these prosperous companies presented a dilemma to gene therapists. On the one hand, they needed investment for what was (and still is) a very expensive science. In contrast, the needs to bring about fast returns and to work on more genetically complex conditions were not always compatible with the limitations of gene therapy.

Researchers also faced considerable difficulties in simply making gene therapy work. Retroviral vectors continued to integrate into patients' chromosomes almost at random, and the transferred genes often failed to show up in the body, or last long if they did. Even the diseases themselves were more complicated than anticipated, as seen in the case of CF. Hype can only coexist with a dearth of results for so long. Toward the end of the decade regulators, journalists, and researchers themselves began to acknowledge that enthusiasm for gene therapy had overridden the reality that this was a difficult science.

Its difficulties were further driven home in 1999 when the first patient died during a gene therapy trial. Jesse Gelsinger had just turned 18 when he travelled to the University of Pennsylvania to take part in a gene therapy trial for OCD (ornithine transcarbamylase deficiency), an inherited disorder that he had lived with all his life, which causes ammonia to circulate in the blood. The trial, run by James Wilson and Steven Raper, was what is called a dose escalation trial. This means that participants are put into groups of three, and given increasing doses of the therapy while the researchers watch for any side effects that might arise as the dose increases. Jesse was in the trio to receive the highest dose. Eighteen hours after receiving the treatment, Jesse began to look unwell. Over the following 70 hours his condition deteriorated dramatically, and he died of a massive inflammatory response to the treatment.[29] His death put an immediate stop to the trial, and to several others that were using similar methods.

Receiving coverage in forums including the Washington Post and the New York Times, Jesse's death increased public anxieties about gene therapy. Investigations into his death identified multiple concerns. Primates had died when the project was still in animal testing phases, but no one had reported this to the regulators. (It was later revealed that this kind of underreporting was rampant in the American gene therapy community.) Jesse's health going into the trial was found to be lacking in areas, suggesting that he should not have been given the highest dose of gene therapy.[30] Furthermore, the bioethicist advising on the trial, Arthur Caplan, had argued that parents were swayed by their children's illness and unable to make reasoned decisions about their participation. He recommended that relatively healthy patients be enrolled instead.[31] Financial conflicts of interest were also raised, and formed part of the Gelsinger family's lawsuit against the researchers. The family also argued that the risks of the trial had not been communicated to them adequately.

The trial and the investigations that followed raised serious questions for gene therapists around the world, and saw regulations tighten to minimise the chances of harm to participants in gene therapy trials. Jesse's death not only prompted this tightening of gene therapy regulations but also highlighted the need to find the right scientific method for the desired medical effect.

Jesse was not the last person to die from gene therapy. In 2002 several patients in France, and later in the United Kingdom, developed leukaemia after receiving gene therapy for X-SCID, an inherited disorder of the immune system.[32] The cause of death was found to be the behaviour of the inserted DNA-vector construct. Upon entering the cells the vector integrated into a segment of the chromosome that was close to an oncogene, a cancer-causing gene, which caused cancer cells to proliferate in several of the children participating. These children were given chemotherapy; one died, but the rest survived, and over subsequent years the disease for which they had been treated improved.[33]

By the late 2000's many of the difficulties that had been experienced with gene therapy in the previous decade had begun to find resolution, helped by new generations of vectors which improved the safety of the therapies. At the same time Asian countries including South Korea, China and Japan began to increase their activity in the field, and in 2003 the world's first gene therapy was approved for market use in China.[34] Called Gendicine, the therapy treats head and neck cancers, but limited English language information has left some EuroAmerican observers sceptical about the quality of the trials that produced it.[35] In 2012 the European Commission approved Glybera, for the treatment of the rare disease lipoprotein lipase deficiency. The approval came after several rejections, and a special request from the European Commission that the treatment be reconsidered for a smaller group of patients. No patient has received the treatment to date, and an attempt to gain approval from US authorities was abandoned in 2015. This might reflect the high price of Glybera, which is $1 million per treatment. It is the most costly drug in the world.

6.4 CURRENT ISSUES IN GENE THERAPY

Today approximately 2000 gene therapy trials have been completed or are underway worldwide. Glybera remains the only gene therapy product approved in Europe. In China two gene therapies have been authorised, Gendicine, approved in 2003 for head and neck cancer, and Oncorine, approved in 2005 for nasopharyngel cancer.[36] In 2011 Russian authorities approved Neovasculgen for the treatment of peripheral arterial disease, and the product has since been approved for use in Ukraine as well.[37] US regulators have not so far approved any gene therapies, but one application is pending review.[38]

The low number of gene therapies on the market is not surprising given the difficulties of making such treatments work. Another obstacle to gene therapies reaching the market is the process for gaining regulatory authorisation. Regulators typically base their approval upon a substantial body of evidence gathered from clinical trials involving large number of patients. This model, while important for testing the statistical significance of the particular effect of a drug, is not particularly suited to gene therapies. One of the major problems with gene therapy is that many of the target diseases are rare. As a consequence it can be impossible to enrol a lot of patients for testing a gene therapy. Added to this is the paucity of knowledge about the natural history of the diseases in question. This makes it difficult for researchers to assess the exact impact gene therapy has on the course of such diseases. Funding is a further barrier to getting gene therapies to market. Not only is running a clinical trial extremely expensive, so too is producing or purchasing vectors, clones and the other pieces that make up a gene therapy.

Gene therapy, however, now could be poised to enter a new phase as a result of the development of new gene editing tools which offer the ability to make more precise genetic changes than before and can modify a whole gene with ease. This includes TALENs (transcription activator-like effector nucleases), ZFNs (zinc finger nucleases) and, most famously, CRISPR (clustered regularly interspaced short palindromic repeats). Such gene-editing tools have been used for research in cells and animals for some years, but they gained international public attention after Chinese scientists reported using CRISPR/Cas9 in human embryos in April 2015. The embryos they modified were non-viable, meaning they would never develop into babies, and they were destroyed shortly after the editing was performed. Nonetheless the experiment provoked much heated debate, focusing mainly on two (frequently conflated) issues. First, whether the technology is suitable for use on humans. Second, whether it is ethically acceptable to genetically alter embryos.

The first question was more or less answered by the results of the experiment itself, which were ultimately poor. Not only was the efficiency of the edits fairly low, but they also managed to alter other components of the genome and the editing was not complete in all embryos. The second question rekindled a debate that has been going on since the earliest days of gene therapy. Opponents of embryonic gene therapy argue that it is subjecting a future individual to an experimental procedure that they cannot consent to. What is at stake here, of course, is what an embryo actually is (a person? an unconscious organism? a mere collection of cells?) and who can tamper with it. This conversation is by no means limited to gene therapy: similar questions have been posed in relation to abortion and IVF.

In December 2015 an international summit held in Washington DC on gene editing proposed a temporary solution to the issue, arguing that gene editing research on embryos that would otherwise be discarded should be allowed to continue, but that no person should result from such experiments. The degree to which this will happen will vary by country. Unlike in the USA, where the NIH has banned the funding of research using embryos, European and Asian research groups are already entering the arena. In February 2016, for example, the UK's Human Fertilisation and Embryology Authority (HFEA) authorised UK researchers to apply CRISPR to human embryos under 14 days old as part of a project to understand embryonic development and the genetic causes of miscarriage.

The fact that UK researchers were granted permission to genetically edit human embryos was probably helped by a decision by the HFEA taken the year before to permit the transfer of DNA found in the mitrochondria of the cell, which is separate from the nucleus, where most DNA resides, and important to generating energy within the cell. Mitochondrial transfer shares many similarities with gene therapy in that it involves replacing the mitochondrial DNA (mtDNA) of a woman carrying mitochondrial disease with that of a healthy donor, so that a woman can have a healthy child that is genetically her own even if she carries mitochondrial disease. In the hearings that preceded this technology's approval much was made of how little mtDNA actually does; critics argue, however, that we know too little about this genetic substance to really say. What is interesting here is that mtDNA transfer is not discussed as a form of gene therapy, despite relying upon the removal and replacement of a portion of one person's DNA with another. The pertinent question is whether only changes made to nuclear DNA are now thought of as being gene therapy.

6.5 CONCLUSION

Debates like those over gene editing and mtDNA science are explicitly political. How technologies are represented in such debates affects how they will be perceived, and probably implemented. The recent advances in gene editing suggest that gene therapy might be due a rethinking. Perhaps the limits of what can be done using gene transfer will soon be expanded. What is certain is that gene therapy is now soundly positioned in certain clinics around the globe, albeit in an experimental capacity. Increasing numbers of people with genetic diseases will find themselves with the option of taking part in gene therapy clinical trials. The hope and risk and available options for such individuals will shift as gene therapy becomes a viable option for them. These patients will live out in real life the philosophical questions that surround the implementation of such therapy. While patients and parents decide whether to try out an experimental gene therapy, questions will continue as to how far gene therapy can go and to whom it should be available—children, adults, embryos?

REFERENCES

1. M. L. Edelstein, M. R. Abedi and J. Wixon, *J. Gene Med.*, 2007, **9**, 833.
2. L. Seymour and A. Thrasher, *Nat. Biotechnol.*, 2012, **30**, 588.
3. N. Cartier, S. Hacein-Bey-Abina, C. C. Bartholomae, G. Veres, M. Schmidt, I. Kutschera, M. Vidaud, U. Abel, L. Dal-Cortivo, L. Caccavelli, N. Mahlaoui, V. Kiermer, D. Mittelstaedt, C. Bellesme, N. Lahlou, F. Lefrère, S. Blanche, M. Audit, E. Payen, P. Leboulch, B. l'Homme, P. Bougnères, C. Von Kalle, A. Fischer, M. Cavazzana-Calvo and P. Aubourg, *Science*, 2009, **5954**, 818.
4. S. Ghoash, A. J. Thrasher and H. B. Gaspar, *Br. J. Haematol.*, 2015, **171**, 155.
5. J. G. Mikkelsen, *Somatic Genome Manipulation*, ed. X. Q. Li, D. J. Donnelly and T. G. Jensen, Springer Science, 2015, ch. 4, p. 69.
6. U. Griesenbach and E. W. F. Alton, *Hum. Mol. Genet.*, 2013, **22**, 52.
7. B. L. Evatt, *J. Thromb. Haemostasis*, 2006, **4**, 1982.
8. N. Comfort, *The Science of Human Perfection: How Genes Became the Heart of American Medicine*, Yale University Press, 2013.
9. J. A. Wolff and J. Lederberg, *Hum. Gene Ther.*, 1994, **5**, 469.
10. M. Nirenberg, *Science*, 1967, **157**, 633.
11. H. Landecker, *Culturing Life: How Cells Became Technologies*, Harvard University Press, 2007.

12. S. Wright, *Molecular Politics: Developing American and British Regulatory Policy for Genetic Engineering, 1972–1982*, University of Chicago Press, 1994.
13. P. Berg, D. Baltimore, H. W. Boyer, S. N. Cohen, R. W. Davis, D. S. Hogness, D. Nathans, R. Roblin, J. D. Watson, S. Weissman and N. D. Zinder, *Science*, 1974, **4148**, 303.
14. D. S. Frederikson, *Asilomar and recombinant DNA*, ed. K. E. Hanna, National Academies Press, 1991, Biomedical Politics, p. 258.
15. W. F. Anderson and J. C. Fletcher, *N. Engl. J. Med.*, 1980, **303**, 1293.
16. H. M. Schmek, *The New York Times*, 1981, U.S. agency disciplines gene therapy researcher.
17. T. Friedman, *Nat. Genet.*, 1992, **2**, 93.
18. C. Randall, B. Mandelbaum and T. Kelly, *Letter from three general secretaries*, 1980, Letter to Congress.
19. President's Commission for the Study of Ethical Problems in Medicine and Biomedical and Behavioral Gene Research, Report, 1982, Washington.
20. W. F. Anderson, *The New Genetics and the Future of Man*, ed. M. Hamilton, Eerdmans, 1970, Genetic Therapy, p. 109.
21. W. F. Anderson, *Hearings on Human Genetic Engineering Before the Subcommittee on Investigations and Oversight of the Committee on Science and Technology*, 1982, Washington, 285.
22. C. P. Addison, *BioSocieties*, 2017, **12**, 257.
23. J. Lyon and P. Gorner, *Altered Fates: Gene Therapy and the Retooling of Human Life*, WW Norton & Company, 1996.
24. S. A. Rosenberg, P. Aebersold, K. Cornetta, A. Kasid, R. A. Morgan, R. Moen, E. M. Karson, M. T. Loetz, J. C. Yang, S. L. Topalian, M. J. Merino, K. Culver, D. Miller, M. Blaese and W. F. Anderson, *N. Engl. J. Med.*, 1990, **323**, 570.
25. R. M. Blaese, K. W. Culver, A. D. Miller, C. S. Carter, T. Fleisher, M. Clerici, G. Shearer, L. Chang, Y. Chiang, P. Tolstoshev, J. J. Greenblatt, S. A. Rosenberg, H. Klein, M. Berger, C. A. Mullen, W. J. Ramsey, L. Muul, R. A. Morgan and W. F. Anderson, *Science*, 1995, **270**, 475.
26. C. Bordignon, F. Mavilio, G. Ferrari, P. Servida, A. G. Ugazio, L. D. Notarangelo, E. Gilboa, S. Rossini, R. J. O'Reilly, C. A. Smith, A. P. Gillio, W. F. Anderson, R. M. Blaese, R. C. Moen and M. A. Eglitis, *Hum. Gene Ther.*, 1993, **4**, 513.

27. J. Lyon and P. Gorner, *Altered Fates: Gene Therapy and the Retooling of Human Life*, WW Norton & Company, 1996.
28. P. Martin, *Sociol. Health Ill.*, 1999, **21**, 517.
29. S. E. Raper, N. Chirmule, F. S. Lee, N. A. Wivel, A. Bagg, G. Gao, J. M. Wilson and M. L. Batshaw, *Mol. Genet. Metab.*, 2003, **80**, 148.
30. J. M. Wilson, *Mol. Genet. Metab.*, 2009, **96**, 151.
31. M. S. Lindee, *Moments of truth in genetic medicine*, JHU Press, 2005.
32. S. Hacein-Bey-Abina, C. Von Kalle, M. Schmidt, M. P. McCormack, N. Wulffraat, P. Leboulch and M. Cavazzana-Calvo, *Science*, 2003, **302**, 415.
33. S. Hacein-Bey-Abina, A. Garrigue, G. P. Wang, J. Soulier, A. Lim, E. Morillon and M. Cavazzana-Calvo, *J. Clin. Invest.*, 2008, **118**, 3132.
34. S. Kim, Z. Peng and Y. Kaneda, *Mol. Ther.*, 2008, **16**, 237.
35. J. M. Wilson, *Hum. Gene Ther.*, 2005, **16**, 1014.
36. L. C. Kaptein, Y. Li and G. Wagemaker, Report, Netherlands Commission on Genetic Modification, 2010.
37. Human Stem Cells Institute, Press Release, Moscow, 2011.
38. Spark Therapeuticsm, Press Release, Paris, 2015.

CHAPTER 7

Stem Cells: An Emerging Field for Medicine[†]

ALISON KRAFT*[a] AND FRANK BARRY[b]

[a] Max Planck Institute for the History of Science, Boltzmannstr. 22, 14195 Berlin, Germany; [b] Regenerative Medicine Institute, Ireland
*Email: akraft@mpiwg-berlin.mpg.de; frank.barry@nuigalway.ie

7.1 INTRODUCTION

In February 2017 the journalist Andrew Marr broadcast a documentary that chronicled his recovery from a stroke that he suffered in 2013. This featured the experiences of fellow patients to illustrate the devastating and diverse effects of stroke, and the different challenges faced by stroke survivors. One such patient was Billy Elder, shown undergoing stem cell therapy at the Queen Elizabeth Hospital in Glasgow. He received two million stem cells injected directly into the right hemisphere of his brain. The aim was to improve his movement on the left side of his body, which had been paralysed by the stroke. Mr Elder's physician explained that the stem cells had the power to repair tissue damage, directly and/or indirectly by triggering repair

---

[†]This chapter draws extensively on ref. 1.

Engineering Health: How Biotechnology Changed Medicine
Edited by Lara V. Marks
© The Royal Society of Chemistry 2018
Published by the Royal Society of Chemistry, www.rsc.org

responses. The hoped for therapeutic effect rested centrally upon the production of new cells. All involved recognised the therapy to be highly experimental. As yet there is no definitive evidence to show that stems cells are an effective treatment for strokes.

This vignette underlines the way in which stem cell therapies engender therapeutic visions even as such therapies remain highly experimental. It captures succinctly the hopes attached to their regenerative powers, which have captured the public and scientific imagination alike. The idea that stem cells might offer a new means to treat a range of prevalent chronic diseases, including stroke and also ischaemic heart disease and numerous degenerative diseases of the brain has, since the late 1990s, spurred commercial interest and investment in stem cells and the vast expansion of stem cell research.

The clinical use of stem cells, however, is not new. As this chapter shows, contemporary developments rest in part on advances arising out of research into the blood stem cell in haematological research from the late nineteenth century onwards and the use since the late 1950s of bone marrow transplantation (BMT) as a means to treat leukaemia.[1] This chapter begins by sketching out the history of the blood stem cell and its clinical use under the aegis of bone marrow transplantation. It then examines the emergence of contemporary stem cell innovation and the commercialisation of stem cell therapies.

As will be shown, this can be broken down into three distinct phases. The first 'wave' occurred between the early 1980s and the mid-late 1990s and was focused on the blood stem cell. Specifically, this sought ways to enhance the efficacy of BMT as a treatment for leukaemia. This involved the development of cell separation/processing techniques based on a newly identified cell surface marker, CD34, specific for a population of stem cells in bone marrow which could be used to enrich the stem cell content of marrow for use in autologous bone marrow transplantation. The second wave, which began in the late-1990s, saw major advances in stem cell science and the emergence of the notion of stem cell 'plasticity'. This period correlated with the rise of regenerative medicine, which underpinned the rapid growth of a new biotechnology sector centred on stem cells. The third wave followed in the wake of the creation in 2006 by a Japanese group of an engineered stem cell, referred to as the induced pluripotent stem (iPS) cell.

7.2 THE 'STEM CELL' CONCEPT

Historically, stem cells have been understood to be the starting point in cell formation, a process which involves cell proliferation (cell division) and cell differentiation. The term stem cell (*Stammzelle*) has been attributed to the German biologist Ernst Haeckal in 1868.[2] He coined the term in the context of evolution, referring to a primitive unicellular organism from which all multicellular organisms evolved. Haeckal and others subsequently adapted the term within embryological research, referring to the fertilised embryo as a stem cell, the cellular entity that gives rise to all the cell types of the body.[3]

By the 1890s the term stem cell had been adopted by those undertaking research into blood cell formation to describe a distinctive kind of cell resident in bone marrow. Understood to be endowed with unique powers of cell proliferation and differentiation, this cell was also understood to function as the source of cell renewal in the blood system. However, this cell proved elusive: for decades it could be neither identified or isolated using available techniques.[4,5]

7.3 HAEMATOPOIETIC STEM CELLS (HSCs)

By the early twentieth century scientists had begun to determine the cellular composition of blood, and were developing a model of the blood system centred on distinct haematopoietic lineages. One of the key figures in this research was Artur Pappenheim, a German physician and pathologist who, between 1896 and 1898, proposed that different blood cells were derived from a common ancestor, which he called the Stammzelle or stem cell (Figure 7.1). The concept of a common stem cell for the blood system is now firmly established within haematological research, and similar models of cell formation have been applied to other populations of stem cells in other tissues. Further and compelling evidence for the tissue-specific multipotential model of the blood stem cell came during the 1960s in work by a Canadian group led by James Till and Ernest McCulloch.[6,7]

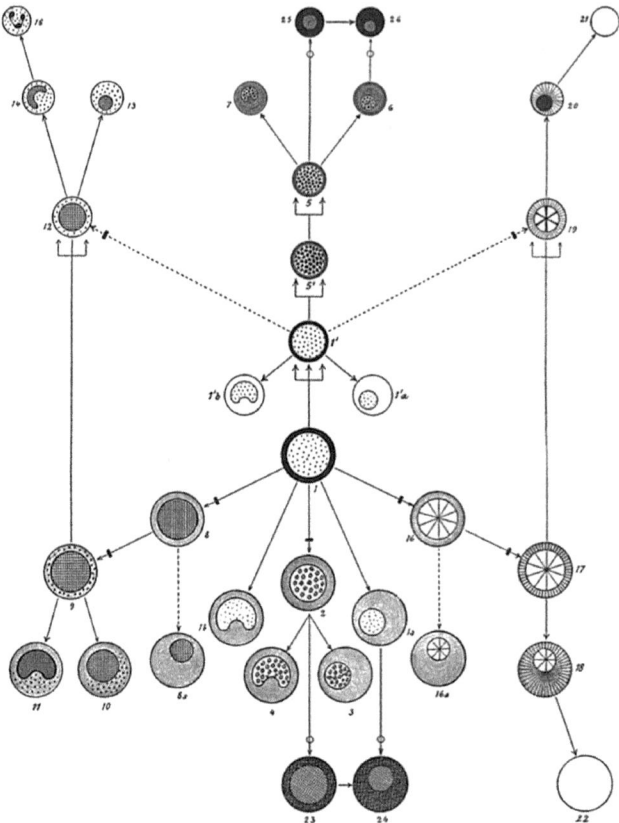

Figure 7.1 Pappenheim's diagram of the production and differentiation of different blood cells. All the blood cells are derived from the one in the centre which Pappenheim called the Stammzelle, or stem cell.[8] Reprinted with permission from *Cell Stem Cell*, Volume 1, M. Ramalho-Santos, H. Willenbring, On the Origin of the Term "Stem Cell", 35–38, Copyright (2007), with permission from Elsevier.

7.4 BONE MARROW TRANSPLANTATION

As early as the 1930s, the blood system was recognised to be acutely sensitive to ionising radiation. By the 1950s this phenomenon had become an intense area for research as a result of the changed scale and scope of radiation hazards in the 'atomic age'. This created a pressing need to understand how and why blood cells and the bone marrow were so susceptible to the damaging effects of ionising radiation.[9]

Out of this work came a new technique called bone marrow transplantation (BMT). Initially used as a research tool for studying the effects of radiation on the blood system in animal models, BMT was soon found to have clinical potential as a new means to treat some forms of leukaemia. BMT is a two-step technique that involves the destruction of the patient's blood system using lethal whole-body irradiation to eradicate leukaemic cells, followed by the regeneration of the patient's blood system by the transplantation of (healthy/histocompatible) donor marrow containing stem cells. The aim is to replace the patient's marrow with healthy stem cells from the donor thereby restoring a healthy immune system and haematopoietic system. In the late 1950s, BMT was radically new and its use in patients the subject of debate.

The first account of the clinical use of BMT was published in September 1957 by a team led by E. Donnall Thomas in a paper reporting their findings in six terminally ill patients. Thomas and others perceived that beyond its use as a treatment for severe blood disorders, BMT might also be useful as a potential therapy for radiation exposure—something which, in the late 1950s at the height of the Cold War, took on new importance.[10]

Significantly, in the late 1960s, BMT was also being used in another clinical context. In 1968, Robert Good and colleagues at the University of Minnesota pioneered its use in immune-deficiency disorders, administering donor marrow to a 5 month old boy suffering from this kind of disease which had already killed several members of his extended family.[11] In this case the donor was the boy's 8 year old sister: the child's immune system was restored and he survived into adulthood.[12]

The early clinical use of BMT as a means to treat leukaemia was fraught with difficulties, including unforeseen and serious immunological reactions, principally 'graft *versus* host' disease (GvHD).[13] Early results were so disappointing that it was abandoned for much of the 1960s. During this decade Thomas and other groups returned to animal models to investigate the complex immunological problems that hindered its use in patients. Clinical trials resumed in the 1970s: the BMT protocol was now a complex multi-step regimen that included immuno-therapy and chemotherapy elements.[14] This improved patient outcomes, so that by the early 1980s BMT was an established part of the therapeutic repertoire for treating some forms of leukaemia.

Further improvements in the clinical efficacy of BMT followed from work published in 1984 by Curt Civin and colleagues at the Johns Hopkins University Medical School which identified a molecular tag—the CD34 surface marker—that was specific for a population of stem cells resident in human bone marrow.[15] These techniques opened up, for the first time, a reliable means of producing marrow rich in stem cells that were cancer-free and safe for the patients' immune systems.[16,17]

Since then many thousands of children and adults have received BMTs for a range of life-threatening conditions, including leukaemia, multiple myeloma, aplastic anaemia, sickle cell anaemia and severe immune deficiency. Although not without serious side effects, including GvHD, the treatment has been a life-saver for many patients, especially paediatric patients.

7.5   STEM CELL THERAPY BEYOND BMT

Early on it was understood that BMT succeeded because stem cells in the healthy donor marrow, engrafted, differentiated and reconstituted a new marrow in the patient. Research into these processes opened a window onto the blood stem cell, and clarified understanding about both blood cell formation and the physiology of the blood and immune systems. By the 1960s the blood stem cell had become the most intensively studied somatic stem cell—a position consolidated by the work of the Till and McCulloch group in Toronto. In effect, the blood stem cell became paradigmatic for adult/somatic stem cells generally, providing the conceptual and scientific framework for analysing and understanding other populations of stem cells.

Stem cells were soon perceived as having therapeutic potential beyond BMT and the set of diseases, notably the leukaemias, with which, to date, this technique had been associated. One of the first to grasp this wider therapeutic potential was the Hungarian-born cell/radiobiologist Laszlo Lajtha, director of the Paterson Institute for Cancer Research in Manchester, UK and, by the 1970s, one of the world's foremost experts on the blood stem cell.[18]

Lajtha envisaged, as early as 1975, that the major challenge to harnessing the therapeutic potential of stem cells lay in controlling cell differentiation. As he put it in 1979, 'We are only beginning to understand some aspects of stem cell control. The one nearest to manipulation by experimental techniques is their proliferation control. The next big ask, still awaiting future methodologies, is manipulation of their differentiation' which, he argued, 'may lead to the prize of purposeful manipulation of cell specialization in the interests of man'.[19,20]

Lajtha's prescient anticipation of the therapeutic potential of stem cells and of the key challenge of controlling cell differentiation has, decades later, begun to move closer to realisation, enabled by myriad scientific advances and by new techniques flowing from biotechnology. Not least, this brought a novel toolbox of powerful techniques for analysing the genetic aspects of cellular life, which was closely bound up with the growing dominance of molecular biology and of genetics-based approaches to the life of the cell. Molecular techniques were used, for example, by the Civin group in 1984 to identify the CD34 surface marker for stem cells within bone marrow which, as noted above, provided the basis for a new wave of stem cell innovation during the 1990s.

7.6 FIRST WAVE OF COMMERCIALISATION: CD34—CELL SEPARATION AND STEM CELL ENRICHMENT

The clinical import of the discovery of CD34 was immediately recognised to have commercial potential. Now, CD34 separation and enrichment protocols could, potentially, provide a means to concentrate the regenerative power of bone marrow with attendant clinical benefits to the patient, including speeding marrow regeneration and the restoration of marrow function. This provided the basis for a wave of commercial interest centred on the niche area of BMT and its use in treating blood disorders, especially leukaemias. By 1993, six companies, most of which were biotechnology 'start-ups' and all of which were US-based, were developing various techniques for the isolation and purification of CD34 cells.[21]

However, CD34 separation was beset with technical difficulties, and a business model based on bespoke cell preparations for individual patients presented serious challenges. Moreover, problems surrounded the creation and delineation of intellectual property. Indeed, CD34 became the focus of numerous intellectual property claims, and the subject of bitter litigation between two of the six 'first wave' companies, Baxter Healthcare Corporation and CellPro.[22]

All of this stymied the growth of the sector which, by the mid-1990s, was in serious difficulty. That said, 'big pharma' was showing some interest in CD34-based enrichment technologies: in 1995, Rhone-Poulenc Rorer (now Aventis Sanofi) acquired AIS for $220 m and in 1997 Sandoz (now Novartis) acquired Systemix for $600 m. Meanwhile, CellPro ceased trading after losing its patent dispute with Baxter. By the mid-late 1990s, the first attempts to render stem cells—specifically the blood stem cell—as a commercial proposition had floundered.

7.7   SECOND WAVE OF COMMERCIALISATION: REGENERATIVE MEDICINE

By the late 1990s, the blood stem cell was fast moving to the forefront of another therapeutic paradigm, that of 'regenerative medicine'. Echoing BMT, this rested on harnessing the inordinate powers of cell proliferation and differentiation with which stem cells were endowed. This new therapeutic vision emerged out of three disparate lines of research, including that culminating in the birth in 1997 of Dolly the sheep, work which showed the possibility of somatic cell nuclear transfer. Secondly, emerging evidence indicated that, in contrast to the historical view of the blood stem cell as tissue-specific, it might be able to give rise to cells other than those of the blood system. Thirdly, the laboratory cultivation of human embryonic stem cells elicited great excitement that biomedicine was on the cusp of a new therapeutic era. All three developments were associated with a new wave of stem cell innovation that differed markedly from that of the early 1980s. Now, recast as regenerative medicine, the horizon of stem cell innovation broadened out to encompass different types of adult stem cell newly envisioned as prospective therapies for a range of chronic, prevalent degenerative diseases.[23]

The science underlying the emerging paradigm of regenerative medicine merits brief elaboration. The birth in Edinburgh in 1997 of the cloned sheep Dolly caught the public imagination and stunned the scientific world.[24] Conceptually, Dolly engendered a profound shift in thinking about the process of cellular differentiation. She demonstrated that under certain conditions the nucleus of a fully differentiated mammalian cell could revert to the earliest embryonic stage, that of the fertilized egg. Dolly was living proof that the differentiated (specialized) adult cell could be 'reset'.[25] Here was an opportunity for manipulating cellular development. Historically, the blood stem cell had been understood to be multipotent but tissue specific, that is it was able to give rise to the different types of cell—red, white and so forth, that constituted the blood system. New findings in the late 1990s, however, began to challenge this idea, indicating that in some circumstances the blood cell stem cells had the capacity to differentiate into a much wider range of cells.[26-28] This phenomenon, referred to scientifically as 'transdifferentiation' and colloquially as 'plasticity', was fiercely contested at the time.[29,30] Nonetheless, it lent further momentum to visions of and research into stem cell therapies. The observation of veteran blood stem cell expert Ihor Lemischka that the suggested plasticity of somatic stem cells 'may revolutionize the way we think about tissue transplantation therapies and regenerative medicine' affords some sense of the way in which 'plasticity' was immediately coupled to a vision of clinical utility.[31,32] As Blau and colleagues observed in 2001, 'the ability of stem cells from multiple sources to regenerate diverse tissues greatly increases the flexibility and applicability of tissue regeneration strategies'.[33] That said, the idea that the blood stem cell could, under some circumstances, give rise to cells other than those of the blood system - a possibility expressed in terms such as 'turning blood into brain', remained deeply controversial. The science remained uncertain.

These developments coincided with another groundbreaking advance: the *in vitro* cultivation of human embryonic stem cells. As we will see, this work,[34] first reported in 1998 by an American group led by James Thomson fuelled further excitement about stem cell therapies. All three lines of stem cell related research combined to lend new momentum to the idea of controlling cell proliferation and differentiation for therapeutic purposes.

7.8   MESENCHYMAL STROMAL CELLS (MSCs) THERAPY

Meanwhile, interest was growing for another group of stem cells known as mesenchymal stromal cells (MSCs). These cells were first described in the 1960s by Friedenstein,[35] who identified a subpopulation of cells within the stromal compartment of bone marrow with osteogenic potential.[36,37] By the late 1980s Friedenstein and his colleagues had shown that these adherent, fibroblastic-like cells were capable of forming colonies from a single cell and had the capacity to form multiple skeletal tissues *in vivo*.[38] Subsequently, in the 1990s, individual, clonal populations of stem cells in human bone marrow were identified which retained the potential to differentiate into chondrocytes, osteocytes and adipocytes.[39]

Since 1999, MSCs have been isolated and characterized from many other human sources including adipose tissue,[40–42] umbilical cord blood,[43,44] and Wharton's jelly.[45,46] They all have the capacity to differentiate into cells of connective tissue lineages, including bone, fat, cartilage and muscle. Bone-marrow-derived MSCs have an additional potential in providing the stromal support system for haematopoietic stem cells.[47–49]

MSCs provided a number of advantageous characteristics for clinical use, including the ease with which they could be isolated and their numbers expanded *in vitro* in the laboratory. Important too were indications that these cells could be used in the allogeneic setting *i.e.* transplantation between immunologically 'mismatched' donors was possible without, seemingly, eliciting serious immunological reactions and/or outright 'rejection' in the recipient.[50]

One of the first times MSCs were clinically used was in 1991. This work was carried out by Al Caplan and his colleagues at Case Western Reserve University. They reported the use of the cells for cartilage repair, a condition for which very few treatment options exist.[51] This provided the basis for the development of the concept of MSC therapy for the treatment of orthopaedic conditions. In 2001 another team reported the use of osteoprogenitor cells isolated from patient bone marrow, expanded *ex vivo*, and placed on macroporous hydroxyapatite scaffolds prior to transplantation in large bone defects of the humerus, tibia or ulna.[52] Their results indicated a rapid repair response in all cases, faster then would be seen with traditional bone allografts. This approach was based on a tissue engineering concept, whereby the delivered cells would contribute to the formation of new tissue, based on their capacity to act as progenitor cells.

Over time, the tissue engineering concept involving MSC delivery was replaced by a newer, more complex and subtle principle (it is important to note a different but also rapidly developing lineage of stem cell innovation arising out of the tissue engineering tradition[53,54]) One of the first reports of this was provided by Murphy and colleagues, who showed, in a series of animal experiments, that delivery of MSCs to the stifle joint of goats with surgically induced osteoarthritis resulted in improvement of cartilage regeneration but without evidence of significant engraftment of the delivered cells. The result indicated that a different mechanism may underlie the therapeutic effects, namely that the delivered cells contributed to a paracrine response whereby they provided repair signals to the host rather than contributing directly to formation of new tissue. This was a new concept and was subsequently confirmed in a new series of human studies, the results of which indicated that MSCs, rather than acting as progenitors for tissue formation, acted as immunomodulatory cells with the potential to treat a variety of autoimmune conditions, such as GvHD.[55,56]

In subsequent years, MSC therapy was proposed and tested as a treatment for an exceptionally wide spectrum of human diseases,[57] often without a clear underlying hypothesis or understanding of mechanism of action. The results of clinical trials to date have been mixed and, although there are many listed in public databases, these are mostly phase 1 and phase 2. There is still very limited phase 3 clinical trial data available, so any definitive or unambiguous interpretation is not yet possible.

In addition to the lack of progress in clinical trials, there is another aspect of MSC technology that has hampered development of the field, namely a perceived lack of homogeneity in product characterization. It has been clear that there is a great deal of variability in terms of donor and tissue source, protocols for expansion and surface characterization. This lack of attention to product characterization has created scientific and regulatory hurdles that have yet to be overcome.[58]

7.9 HUMAN EMBRYONIC STEM CELLS

Arguably the most controversial area of stem cell innovation is that based on the human embryonic stem cell (hESC) which, since the 1990s, has likewise engendered enormous hopes regarding its therapeutic promise. Crucial here was work by a Madison, Wisconsin-based group led by James Thomson, which in 1998 reported the successful isolation and cultivation of human embryonic stem cells (hESC). The hESC was understood to be totipotent and, as such, capable of giving rise to any and all types of cell found in the body. This was quickly seized upon as a means for providing an unlimited source of specific cells types for therapeutic use. Yet, equally quickly, this provoked concern because it involved the harvesting of cells from preimplantation embryos. Controversy pivoted about the question as to how to protect the embryo whilst also using it as a source of hESCs with a view to the prospective use of these cells as novel therapies for patients with severe and intractable medical conditions. This engendered an intense and protracted debate about the ethical sourcing of stem cells for therapeutic use, which resulted in the application of restrictions in some countries, and, in all cases, a very high level of regulatory oversight.[59,60]

Where hESC therapy sparked considerable interest was in the repair of spinal cord injuries and for the treatment of neurodegenerative diseases such as Alzheimer's and Parkinson's disease. Amongst the early entrants into the development of hESC therapy for acute spinal cord injury was the California company Geron, which sponsored the first patient study of oligodendrocyte progenitor cells derived from hESCs.[25] Five patients were treated and the results indicated no positive benefit in terms of neurological or sensory restoration. This study, although preliminary, had a negative commercial outcome for the company and highlighted the hazards of carrying out early clinical trials in a field hampered by ethical issues and overly optimistic expectations.[61]

Nevertheless, the totipotency of the hESC, together with the means for cultivating these cells in the laboratory, remains a compelling therapeutic prospect. Research is therefore continuing. Currently, for example, one promising line of work focuses on the use of hESCs or retinal pigment epithelial cells derived from hESCs for the treatment of macular degeneration of the retina where early results show some promise. Chapter 8 offers more detailed analysis of the current status of this area of stem cell innovation.[62,63]

7.10   A GROWING FIELD

The late 1990s to 2006 proved an exciting if turbulent period within stem cell biology. Journals provide one barometer or index of the spectacular growth of stem cell research in the wake of developments such as Dolly, the lab cultivation of the hESC and reports of stem cell plasticity. For example, subscription and submission to *Stem Cells*, first published in 1981, increased markedly and its 'impact factor rating' soared. The readership of the journal's on-line edition, launched in 2000, rose from an initial 2000 per week to over 80 000 per week just two years later. By January 2007 it stood at 120 000 per week.[64,65]

Some sense of the rapid rise to prominence of the field is apparent in the portrayal in 1999 by the influential US journal, *Science* of stem cells as the 'breakthrough of the year'.[66] Around this time, stem cell therapies were assuming new importance within national healthcare plans and strategies, which in policy discourse were often framed in terms of 'regenerative medicine'.[67] Privately and state-funded institutions dedicated to the development of stem cell therapies sprang up around the world, including in America the California Institute of Regenerative Medicine established in 2004, which by 2012 had channelled $1.5 bn into stem cell research.

By 2006 the commercial development of stem cells had undergone considerable growth and segmentation. One survey carried out by Martin and colleagues in 2006 identified a total of 166 stem cell companies, 59% of which were based in the USA, 28% in Europe, with the remaining 13% (approximately 33 companies) distributed fairly evenly between Asia, Australasia, Africa, South America and the Middle East.[68]

7.11   UMBILICAL CORD BANKING

By far the most dominant sector was umbilical cord blood (UCB) banking, which accounted for 35% of the companies surveyed in 2006. UCB banking stood firmly in the tradition of BMT and the blood stem cell, on which, conceptually and scientifically, the banking and prospective use of cord blood rests. UCB had long been known to be rich in stem cells. The first UCB transplant took place in 1989 when it was used successfully to treat a child with Fanconi's anaemia.[69]

The unique biological properties of stem cells present in umbilical cord blood, especially their immunological naivety which, in the clinic, lessened the chances of severe immunological complications, formed the basis of a new business sector centred on harvesting these cells at birth and storing them frozen at very low temperatures in a 'bank'. Uniquely available only at birth, these stem cells were now held in perpetual readiness, mobilisable should the child subsequently become ill with any of the several diseases for which UCB was indicated. At this point, the stem cells present in cord blood and hitherto stored in a 'bank' were considered potentially of therapeutic use.[70]

The first UCB banks, both public and private, were formed in 1992 in the US, with a business model based on a one-off fee for initial collection (approximately $1000) and a smaller annual fee for storage. The period 1997 to 2004 saw rapid growth in UCB banking, with the number of UCB banks rising from 32 in 1997 to 119 in 2004.[71] Much of this was spurred on by the new ideas of stem cell plasticity and regenerative medicine. The commercial success related more to its 'banking' role and its status as offering an (uncertain) insurance against future illness, rather then the actual clinical use of the stored stem cells.[72]

Both American and British companies continued to dominate, accounting for 29% and 24% of the sector respectively: the five largest banks were located in either the USA or the UK, which, together, held 70% of all UCB. However, by 2004, UCB banks had been established in China, Japan, Singapore and the Middle East. By this time, 66% of UCB banks were commercially operated and their holdings were for family/private use: technically, cord blood stored here was principally for autologous use. The remaining 33% of UCB banks were publicly funded and based on donation, cord blood stored in this way was premised on altruism and geared for clinical use in the allogeneic setting.[73] The upswing in UCB banking in the opening years of the 21st Century was directly linked to the ascendency of stem cell plasticity and to the promise of stem-cell therapies. As Martin *et al.* note, UCB banks rested on 'the promise of stem-cell-based regenerative medicine' and formed part of what they call the 'promissory bioeconomy'.[71,74,75]

7.12 STEM CELL INNOVATION: THE OVERALL LANDSCAPE

Looking beyond UCB banking, the picture was more complicated and less commercially encouraging. Companies working with adult stem cells, representing some 43% of the sector, were pursuing diverse strategies, with many focused on cell processing and purification, encompassing stem cell growth factors, tools and reagents linked to handling, separating and growing stem cells. The website of Aldagen, founded in 1999, exemplified the promissory visions of the period, asserting that, 'The use of adult stem cells for the purpose of tissue regeneration offers the opportunity to revolutionize treatment of a broad range of serious diseases affecting millions of patients'. By 2006, companies working with adult stem cells of various kinds, including blood/bone marrow, neural, skin and muscle, accounted for 43% of the stem cell sector. For all the technical difficulties and ethical controversies surrounding the hESC, some 14% of the sector was working on developing therapies using this cell.[68] Segmentation was not, however, an indication of the strength or stability of the sector.

The continuing dominance of the blood stem cell was also apparent beyond UCB banking. Aastrom Biosciences sought to use bone marrow-derived adult stem cells for bone, cardiac and neural regeneration. In 2006, UCB banking and the category of 'bone marrow stem cells' accounted for 46% of stem cell companies. It is perhaps not surprising that the blood stem cell lay in the vanguard of taking plasticity and regenerative therapies into the clinic. This step was taken first in the USA, where by the end of 2006, some 40 or so trials had been completed or were under way assessing the efficacy of bone marrow stem cells as a treatment for coronary heart disease, heart failure and myocardial infarction. A number of trials were also being initiated to explore the use of HSCs in other degenerative conditions, including Parkinson's disease. However, these early trials yielded mixed and, for the most part, disappointing results and led to criticisms about whether the move into the clinic had been premature.[76,77] Even taking the blood stem cell beyond the province of cancer and BMT was looking decidedly unpromising and dimmed further the outlook for clinical trials of other populations of stem cells not yet shown to be therapeutically usable, let alone useful.

7.13   HURDLES TO COMMERCIALISATION

Delivering on the 'stem cell promise' was checked by complex biology. Stem cell science, in the sense of understanding plasticity, pluripotency and cell differentiation, remained at a very early stage. Harnessing the proliferative and differentiation powers of stem cells for therapeutic benefit was entering into uncharted territory and laden with daunting scientific, technical and clinical challenges. For example, the very small number of stem cells that could typically be isolated from adult human tissue was for some the 'biggest technical hurdle' to the development of stem cell therapies. Others worried that stem cells had a 'dark side', being implicated in the aetiology of some cancers, *via* the theory of the cancer stem cell.[5,78,79] The field was also damaged by high-profile controversies. One of the most well publicised was that surrounding the South Korean researcher Woo Suk Hwang who, in 2004, was dismissed from his post after fabricating evidence that he had successfully cloned hESCs.

Pragmatically, a more realistic view of the timescale settled over the sector. However this was unpalatable to commercial investors. Industry commentators spoke of stem cells as a 'hard sell' to investors, citing variously the disappointing commercial history of stem cells, ethical issues around hESCs and technical difficulties with adult/somatic stem cells, for example, there remained no conclusive evidence that blood stem cells could contribute to the regeneration of tissue beyond the blood and immune system.[80]

Stem cell innovation was also blighted by enduring problems in the creation and delineation of intellectual property. Equally, 'big pharma' remained reluctant to engage with stem cell biology and innovation, adopting, instead, a strategy of 'wait and see'. All of this made it difficult for commercial actors to raise funds from private equity and venture capital. By 2006, the vision of stem-cell-based regenerative therapies based on the principle of 'plasticity' was floundering.

Practitioners and commentators alike were agreed that moving forward with the development of stem cell therapies required closer collaboration between academic researchers, clinicians and commercial actors *i.e.* those working in the laboratories of biotech and pharmaceutical companies. Stem cell therapies called for and exemplified translational research which could transcend the divisions arising from the organizational structures of late twentieth century biomedical research, including institutional barriers and the thorny problem of sharing data and its corollary, intellectual property.[81]

Cell-based therapies also called for new approaches to the regulatory framework governing pharmaceutical innovation which, for historical reasons, was tailored to the dominant paradigm of small-molecule drugs and, since the late 1970s, biopharmaceuticals. Cell-based therapies marked a radical departure from conventional drugs: making changes to the innovation system so as to facilitate their development would take time.[82,83]

Taking stem cells into clinical trials was an especially painstaking and slow process, scientifically, clinically and administratively, and led to a cyclical problem in which the lack of clinical data became a major impediment to moving from bench to bedside. The field also faced setbacks from disappointing results arising from clinical trials that did take place, and here the use of stem cells to treat a range of cardiac conditions featured prominently; trials which had to be abandoned were especially damaging.[84]

Nevertheless, the vision of stem cell therapies proved compelling and research continued apace. Early experiences in treating cardiac diseases with stem cells contributed to a second and highly complex therapeutic strategy which places importance on the stem cell environment *in vivo* and the cellular neighbourhood in the immediate vicinity of disease *e.g.*: the damage caused by ischaemia/infarction. This approach seeks to 'coax' stem cells and other populations of cells in this neighbourhood—the cardiosphere—to initiate repair mechanisms. Several clinical trials were launched seeking to assess and exploit the various therapeutic effects of adult somatic stem cells, including a large, London-based trial exploring the clinical use/efficacy of bone marrow cells in acute myocardial infarction (BAMI).

7.14 THIRD WAVE OF COMMERCIALISATION: THE IPS CELL, 2006–2016

In 2006 a new era in stem cell therapeutics opened up following the creation of a genetically engineered stem cell developed in Japan by a team led by Kazutoshi Takahashi and Shinya Yamanaka.[85,86]

They were part of an international translational research collaboration focused on stem cells. Their findings marked another
unexpected twist in stem cell science. The Japanese scientists had
found a way of manipulating the regulatory mechanisms controlling cell differentiation to 'reprogram' fully differentiated somatic
cells (fibroblasts) to a stem cell state. Likening this genetically altered cell to the embryonic stem cell, Yamanaka and Takahashi
called it the induced pluripotent stem (iPS) cell. The experiments
showed that the introduction of a particular combination of just
four transcription factors[‡] into the cellular milieu was sufficient to
transform the basic 'state' of the differentiated cell (here, the adult
human fibroblast), into that of the iPS cell—with powers approximating those of the embryonic stem cell.

In technical terms, the iPS cell provided a proof of principle
that it was possible to engineer in somatic cells the 'state' of
pluripotency, that is to say, the property of 'stem-ness'. Simply
summarised, iPS cells were made possible by evolving understanding of functional aspects of the genetic circuitry of the cell
and by technical advances in manipulating genetic material. The
new found ability to engineer the pluripotent stem cell state,
embodied in the iPS cell, opened a new window onto the process
of cell formation—that is to say, cell proliferation and
differentiation—and was seen to offer new pathways by which
control might be exercised over these processes for therapeutic
purposes. Laszlo Lajtha's vision of the 1970s was, perhaps,
moving closer.

The iPS cell opened a new chapter in stem cell innovation,
providing a new means for the production of stem cells. This
transformed the availability of stem cells for both clinical uses
and research applications. It also provided the possibility of
generating any of the many different cell types found within one
organism. This was perceived to be a major step towards exer-
cising precise control over cellular differentiation and to represent a 'seismic shift' in stem cell research.[87]

Especially critical, in terms of potential clinical applications,
was the way in which iPS cells could be derived from the somatic
cells of the individual organism and would therefore share its

---

[‡]Specifically, these were: OCT4, SOX2, KLF4 and c-MYC, and were introduced using viral
vectors.

highly specific immunological imprint. As a regenerative therapy or strategy, iPS cells provided a means to circumvent the immunological barrier and its deleterious clinical effects, which had long bedevilled organ and bone marrow transplantation. As the Japanese team emphasised, 'successful reprogramming of differentiated human somatic cells into a pluripotent state would allow the creation of patient- and disease-specific stem cells'.[86]

IPS cells offered several important advantages over regenerative strategies based on both the hESC and adult somatic stem cells, the development of which was, in any case, by this time stymied by various ethical, scientific, technical and political issues. As Gottweis and Minger put it, 'In short, iPS cell research shows promise for a broad range of stakeholders: for stem cell researchers, it is a scientific breakthrough that opens new avenues for regenerative medicine; for the principled opponents of hESC research, iPS cells confirm what they have argued all the while, namely that adult stem cell research was the only way to go; and for policy-makers, iPS cells signify the end of an inconvenient political quarrel with religious fundamentalists and pro-life groups.'[88]

Since 2006, the iPS cell has been the basis for a third wave of commercial interest and investment in stem cell innovation (concerns surfaced early on within iPS cell research that, as with their somatic and embryonic ('natural') counterparts, that the iPS cell might also have a 'dark side' – deleterious mutations and new forms of immunogenicity). As with the second wave, the target indications—and markets—remained prevalent, chronic, degenerative diseases of aging for which, as in the mid-1990s, there remained a dearth of effective therapies. In one sense, iPS cells constituted 'new wine in old bottles'.

Eye diseases leading to blindness, including age-related conditions such as macular degeneration which is the leading cause of blindness in people aged over 65, have been an early and intensive 'hot spot' for iPS cell therapies. This is partly because of practical considerations, including the small number of cells needed, easy surgical access, straightforward assessment of grafts and the practical advantage that one eye can be used as a control, since disease is usually bilateral. To date, the approach has focused on creating iPS cells from the patient.

This is done by growing sheets of tissue *in vitro* for autologous transplantation back into the patient.[89,90] This strategy minimises immunological complications which have long bedevilled corneal transplantation. In 2014, a team based at the Riken Centre for Developmental Biology led by Takahashi began clinical trials with iPS cells made from the skin of a patient with age-related macular degeneration. This involved growing sheets of the retinal pigment epithelium *in vitro* that were then implanted into the right eye of the 70 year old female patient. The results were mixed: the patient reported brightened vision, but the trial was halted following the discovery of two small genetic changes in the eye – which, whilst not implicated in tumour formation, required investigation. Currently, further clinical studies of iPS cell based therapies for this indication are underway.

A second emerging innovation 'hotspot' for iPS cells includes their use as research tools in pharmaceutical innovation, specifically in the drug discovery and development process.[91,92] These applications fall into two broad categories. First, the use of human/patient derived iPS cells, derived and grown *in vitro*, as an *in vitro* screening tool for identifying/optimising drug candidates and assessing toxicity. Secondly, iPS cells offer a new means for *in vitro* modelling of human diseases. Essentially a 'disease-in-a-dish' approach, this line of work uses patient-derived iPS cells, which manifest the cellular/molecular phenotypic markers of disease. Analysis of these cells can shed light on disease mechanisms at a cellular/molecular level and help identify novel drug targets. Advocates argue that as human cell-based assays, iPS cells constitute a 'more physiological system' and have the advantage of approximating human/disease more closely than animal models with attendant benefits for developing effective therapies.

Large pharmaceutical houses, including BMS, GSK, Novartis, Pfizer, Roche and Takeda are particularly interested in this approach and are investing increasing resources into iPS cells. Today there is considerable merger and acquisition activity as large pharmaceutical companies seek to rapidly acquire iPS technology and expertise from smaller biotechnology companies, which, along with institutions given to 'translational' research, formed the vanguard of iPS innovation. iPS cell-derived 'disease-in-a-dish' models have recently 'propelled' neurological drugs into clinical trials at GSK, Roche and BMS. At BMS, the product in development is a treatment for progressive supra-nuclear palsy, currently in Phase I trials, obtained *via* the ac-quisition in 2014 for \$725 m of biotech start-up company iPierian.[92,93]

Combined with gene editing technologies, iPs cells have rap-idly come to be regarded as a 'lab workhorse' providing an 'un-limited supply of once-inaccessible human tissue' for research within and beyond the pharmaceutical setting. Some envisage that iPS cells could radically change the drug development pro-cess, helping curtail pre-clinical and animal studies, with these elements being replaced with *in vitro* clinical trials *i.e.* patient-derived iPS-cell based 'disease-in-a-dish' studies. Some industry and scientific commentators now see the use of iPS cells in drug discovery and disease modelling as eclipsing iPS-based cell therapies, which, as we have seen, is beset with daunting tech-nical challenges.[94–96]

7.15 CONCLUSION

In October 2012, Shinya Yamanaka and John B. Gurdon were awarded the Nobel Prize for Physiology or Medicine for research carried out respectively in the 1960s and in the opening years of the 21st Century for work that established the means for 'reprogramming' mature cells to become pluripotent. As an in-dicator of what, in science, is held to be important, the 2012 Nobel acknowledges the significance of stem cell biology—a field not yet circumscribed in the 1960s, but which by the opening years of the 21st Century occupied the vanguard of biomedical research.

Some idea of the history of the development of stem cells can be seen from Figure 7.2. Each wave of stem cell innovation has been characterised by setbacks, failures, controversies and criticisms, yet the vision of stem cell therapies has remained compelling. In part, interest and investment has been sustained by the continuing failure of either conventional pharmaceuticals (small molecule) or biopharmaceuticals (including monoclonal antibodies) to deliver effective therapies for prevalent chronic degenerative diseases. It has been driven, to some extent, by the

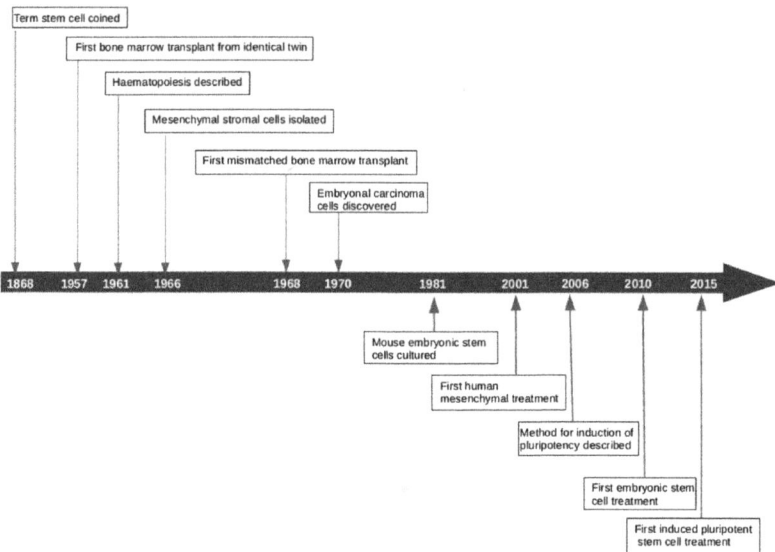

Figure 7.2   Timeline of key developments in stem cell therapy.

'translational imperative' which privileges, prioritises and even demands of biomedical research that investment in it results in clinical and/or commercial applications. Also, it has been forged within a context of ever closer and denser relationships and networks between academic research, clinical practitioners and commercial actors, which are redefining the contours of the biomedical research enterprise.

Embedded within Yamanaka and Gurdon's Nobel Prize are retrospective and prospective elements. The award acknowledges the long pedigree of research into stem cells and the centrality in particular of understanding cell differentiation for the future of stem cell therapies. The Nobel inherently anticipates these therapies. The road to clinical success will probably be a long, slow and frustrating journey: the history of BMT and its continuing value to contemporary stem cell innovation underlines this point. The journey calls for ever closer collaboration between academic, clinical and commercial actors and demands changes to the regulatory framework governing pharmaceutical innovation so as to accommodate novel stem-cell-based therapeutics. Above all, since the 1990s, the goals of stem cell innovation have shifted, but the central challenge—that of controlling cell proliferation and differentiation, within the human body, over a sustained period—remains the same. At the same time, research geared to the development of stem cell therapies of any kind is driving changes to the regulatory framework and forcing a rethink of production models established for conventional drugs.

REFERENCES

1. A. Kraft, *Blood, Radiation, Regeneration: The Elusive Stem Cell c. 1896–c. 1975*, Routledge, Manuscript in preparation.
2. E. Haeckel, *Natürliche Schöpfungsgeschichte*, Georg Reimer, 1868.
3. E. Haeckel, *Anthropogenie*, Wilhelm Engelmann, 1st edn, 1874.
4. For a review of the concept see M. Ramalho-Santos and H. Willenbring, *Cell Stem Cell*, 2007, **1**, 35.
5. H. Maehle, *Notes Records Roy. Soc. Lond.*, 2011, **65**, 359.
6. J. E. Till and E. A. McCulloch, *Radiat. Res.*, 1961, **14**, 213.
7. A. J. Becker, E. A. McCulloch and J. E. Till, *Nature*, 1963, **197**, 452.
8. A. Pappenheim, *Folia Haematol.*, 1908, **6**, 217–242.
9. A. Kraft, *Hist. Stud. Nat. Sci.*, 2009, **39**(2), 171–218.
10. E. D. Thomas, H. L. Lochte, W. C. Lu and J. W. Ferrebee, *N. Engl. J. Med.*, 1957, **257**, 491–496.
11. R. Storb, *Nature*, 2012, **491**, 334.
12. R. A. Gatti, H. J. Meuwissen, H. D. Allen, R. Hong and R. A. Good, *Lancet*, 1968, **28**, 1366.
13. M.-T. Little and R. Storb, *Nat. Rev. Cancer*, 2002, **2**, 231–238.
14. E. D. Thomas, C. D. Buckner, M. Banaji, R. A. Clift, A. Fefer, A. Neiman and R. Storb, *Leuk. Res.*, 1977, **1**, 67.
15. C. I. Civin, L. C. Strauss, C. Brovall, M. J. Fackler, J. F. Schwartz and J. H. Shaper, *J. Immunol.*, 1984, **133**(1), 157–165.
16. Johns Hopkins University, *The Gazette Online*, Aug 16 1999, **28**(42), http://pages.jh.edu/~gazette/1999/aug1699/16invent. html. Accessed Mar 2017.
17. C. Civin, *Lemels N-MIT*, http://lemelson.mit.edu/resources/ curt-civin. Accessed Apr 2017.
18. A. Kraft and B. Rubin, *BioSocieties*, 2016, **11**(4), 497–525.

19. L. Lajtha, *Br. J. Haematol.*, 1975, **29**, 529–537.
20. L. Lajtha, *Differentiation*, 1979, **14**, 23–33.
21. These companies were: AIS, CellPro, Systemix, Aastrom Biosciences, Amcell and Progenitor. For more detailed analysis see ref. 68.
22. A. Bar-Shalom and R. Cook-Deegan, *Milbank Q.*, 2002, **80**(4), 637–676.
23. T. Graf, *Cell Stem Cell*, 2011, **9**(6), 504–516.
24. I. Wilmut, A. E. Schnieke, J. McWhir, A. J. Kind and K. H. Campbell, *Nature*, 1997, **385**, 810–813.
25. S. Franklin, *Health*, 2001, **5**(3), 335–354.
26. C. R. Bjornson, R. L. Rietze, B. A. Reynolds, M. C. Magli and A. L. Vescov, *Science*, 1999, **283**, 534–536.
27. É. Mezey, K. S. Chandross, G. Harta, R. A. Maki and S. R. McKercher, *Science*, 2000, **290**, 1779–1782.
28. D. S. Krause, N. D. Theise, M. I. Collector, O. Henegariu, S. Hwang, R. Gardner, S. Neutzel and S. J. Sharkis, *Cell*, 2001, **105**(3), 369–377.
29. N. Theise and D. Krause, *Leukemia*, 2002, **16**, 542–548.
30. N. Theise, *Exp. Hematol.*, 2010, **38**, 529–539.
31. L. Lemischka, *Exp. Hematol.*, 2002, **30**, 848–852.
32. L. Lemischka, *Proc. Natl. Acad. Sci. U. S. A.*, 1999, **283**, 14193–14195.
33. H. M. Blau, T. R. Brazelton and J. M. Weimann, *Cell*, 2001, **105**, 829–841.
34. J. A. Thomson, J. Itskovitz-Eldor, S. S. Shapiro, M. A. Waknitz, J. J. Swiergiel, V. S. Marshall and J. M. Jones, *Science*, 1998, **282**, 1145–1147.
35. A. J. Friedenstein, I. I. Piatetzky-Shapiro and K. V. Petrakova, *J. Embryol. Exp. Morphol.*, 1966, **16**, 381.
36. A. J. Friedenstein, R. K. Chailakhjan and K. S. Lalykina, *Cell Tissue Kinet.*, 1970, **3**, 393.
37. A. J. Friedenstein, R. K. Chailakhyan, N. V. Latsinik, A. F. Panasyuk and I. V. Keiliss-Borok, *Transplantation*, 1974, **17**, 331.
38. A. J. Friedenstein, *Calcif. Tissue Int.*, 1995, **56**(Suppl 1), S17; M. Owen, *J. Cell Sci., Suppl.*, 1988, **10**, 63.
39. M. F. Pittenger, A. M. Mackay, S. C. Beck, R. K. Jaiswal, R. Douglas, J. D. Mosca, M. A. Moorman, D. W. Simonetti, S. Craig and D. R. Marshak, *Science*, 1999, **284**, 143.

40. P. A. Zuk, M. Zhu, H. Mizuno, J. Huang, J. W. Futrell, A. J. Katz, P. Benhaim, H. P. Lorenz and M. H. Hedrick, *Tissue Eng.*, 2001, 7, 211.
41. J. M. Gimble, A. J. Katz and B. A. Bunnell, *Circ. Res.*, 2007, **100**, 1249.
42. B. A. Bunnell, B. T. Estes, F. Guilak and J. M. Gimble, *Methods Mol. Biol.*, 2008, **456**, 155–271.
43. M. L. Weiss and D. L. Troyer, *Stem Cell Rev.*, 2006, 2, 155.
44. A. Flynn, F. Barry and T. O'Brien, *Cytotherapy*, 2007, **9**, 717.
45. M. L. Weiss, S. Medicetty, A. R. Bledsoe, R. S. Rachakatla, M. Choi, S. Merchav, Y. Luo, M. S. Rao, G. Velagaleti and D. Troyer, *Stem Cells*, 2006, **24**, 781.
46. D. L. Troyer and M. L. Weiss, *Stem Cells*, 2008, **26**, 591.
47. F. P. Barry and J. M. Murphy, *Int. J. Biochem. Cell Biol.*, 2004, **36**, 568.
48. B. Delorme and P. Charbord, *Methods Mol. Med.*, 2007, **140**, 67.
49. P. Bianco, P. G. Robey and P. J. Simmons, *Cell Stem Cell*, 2008, **2**, 313.
50. R. S. Y. Wong, *J. Biomed. Biotechnol.*, 2011, **2011**, 1–8.
51. J. Goshima, V. M. Goldberg and A. I. Caplan, *Clin. Orthop. Relat. Res.*, 1991, **269**, 274.
52. R. Quarto, M. Mastrogiacomo, R. Cancedda, S. M. Kutepov, V. Mukhachev, A. Lavroukov, E. Kon and M. Marcacci, *N. Engl. J. Med.*, 2001, **344**, 385.
53. M. J. Lysaght and A. L. Hazlehurst, *Tissue Eng.*, 2004, **10**(1–2), 309–320.
54. M. Morrison, *Biosocieties*, 2012, 7, 3–22.
55. K. Le Blanc, I. Rasmusson, B. Sundberg, C. Götherström, M. Hassan, M. Uzunel and O. Ringdén, *Lancet*, 2004, **363**, 1439.
56. H. M. Lazarus, O. N. Koc, S. M. Devine, P. Curtin, R. T. Maziarz, H. K. Holland, E. J. Shpall, P. McCarthy, K. Atkinson, B. W. Cooper, S. L. Gerson, M. J. Laughlin, F. R. Loberiza Jr, A. B. Moseley and A. Bacigalupo, *Biol. Blood Marrow Transplant.*, 2005, **11**, 389.
57. A. Tyndall, *Nat. Rev. Rheumatol.*, 2014, **10**, 117.
58. M. Mendicino, A. M. Bailey, K. Wonnacott, R. K. Puri and S. R. Bauer, *Cell Stem Cell*, 2014, **14**, 141.

59. S. Jasanoff, *Designs on Nature: Science and Democracy in Europe and the US*, Princeton University Press, 2011.
60. B. Salter and C. Salter, *Sci., Technol. Human Values*, 2007, **32**(5), 554–581.
61. M. Baker, *Nature*, 2011, **479**(7374).
62. S. D. Schwartz, C. D. Regillo, B. L. Lam, D. Eliott, P. J. Rosenfeld, N. Z. Gregori, J.-P. Hubschman, J. L. Davis, G. Heilwell, M. Spirn, J. Maguire, R. Gay, J. Bateman, R. M. Ostrick, D. Morris, M. Vincent, E. Anglade, L. V. Del Priore and R. Lanza, *Lancet*, 2015, **385**, 509.
63. W. K. Song, K. M. Park, H. J. Kim, J. H. Lee, J. Choi, S. Y. Chong, S. H. Shim, L. V. Del Priore and R. Lanza, *Stem Cell Rep.*, 2015, **4**, 860.
64. C. Civin and A. M. Gewirtz, *Stem Cells*, 2002, **20**, 1–2.
65. L. A. Solberg, *Stem Cells*, 2002, **25**, 847.
66. G. Vogel, *Science*, Dec 1999, 16, http://www.sciencemag.org/news/1999/12/stem-cells-named-breakthrough-year.
67. UK Stem Cell Initiative (November 2005) Report and Recommendations (The Pattison Report), Department of Health.
68. P. A. Martin, C. Coveney, A. Kraft, N. Brown and P. Bath, *Regener. Med.*, 2006, **1**(6), 801–807.
69. E. Gluckman, H. A. Broxmeyer, A. D. Auerbach, H. S. Friedman, G. W. Douglas, A. Devergi, H. Esperou, D. Thierry, G. Socie, P. Lehn, S. Cooper, D. English, J. Kurtzberg, J. Bard and E. A. Boyse, *N. Engl. J. Med.*, 1989, **321**(17), 1174–1178.
70. N. Brown, A. Kraft and P. Martin, *BioSocieties*, 2006, **1**(3), 329–348.
71. P. Martin, N. Brown and A. Turner, *New Genet. Soc.*, 2008, **27**(2), 127–143.
72. N. Brown and A. Kraft, *Technol. Anal. Strategic Manage.*, 2006, **18**(3–4), 313–327.
73. H. W. Busby, Reassessing the 'gift relationship': the meaning and ethics of blood donation for genetic research in the UK, PhD thesis, Nottingham University, 2004, http://eprints.nottingham.ac.uk/10192/1/busby_thesis_final.pdf.
74. B. P. Rubin, *Sci. Cult.*, 2008, **17**, 13–27.
75. B. P. Rubin, *BioSocieties*, 2009, **4**(4), 407–424.
76. K. R. Chien, *Nature*, 2004, **428**, 607–608.

77. R. Passier, L. W. van Laake and C. L. Mummery, *Nature*, 2008, **453**(193), 322–329.
78. R. Lanza and N. Rosenthal, *Sci. Am.*, 2004, **290**, 94–99.
79. A. Kraft, *BioSocieties*, 2011, **6**(2), 195–216.
80. L. B. Giebel, *Nat. Biotechnol.*, 2005, **23**, 798–800.
81. For a clinician view, see: R. Passier, L. W. van Laake and C. L. Mummery, *Nature*, 2008, **453**(193), 322–329.
82. C. Hauskeller and S. Weber, *New Genet. Soc.*, 2011, **30**(4), 415–431.
83. M. S. Corbett, A. Webster, R. Hawkins and N. Woolacott, *BMC Med.*, 2017, **15**, 1–22.
84. A. Abbott, *Nature*, 2014, **509**, 15–16.
85. K. Takahashi and S. Yamanaka, *Cell*, 2006, **126**, 663–676.
86. K. Takahashi, K. Tanabe, M. Ohnuki, M. Narita and T. Ichisaka, *Cell*, 2007, **131**(5), 861–872.
87. G. Holden and C. Vogel, *Science*, 2008, **321**, 756–757.
88. H. Gottweis and S. Minger, *Nat. Biotechnol.*, 2008, **26**, 271.
89. R. Hayashi, Y. Ishikawa, Y. Sasamoto, R. Katori, N. Nomura, T. Ichikawa, S. Araki, T. Soma, S. Kawasaki, K. Sekiguchi, A. J. Quantock, M. Tsujikawa and K. Nishida, *Nature*, 2016, **531**, 376–380.
90. H. Lin, H. Ouyang, J. Zhu, S. Huang, Z. Liu, S. Chen, G. Cao, G. Li, R. A. J. Signer, Y. Xu, C. Chung, Y. Zhang, D. Lin, S. Patel, F. Wu, H. Cai, J. Hou, C. Wen, M. Jafari, X. Liu, L. Luo, J. Zhu, A. Qiu, R. Hou, B. Chen, J. Chen, D. Granet, C. Heichel, F. Shang, X. Li, M. Krawczyk, D. Skowronska-Krawczyk, Y. Wang, W. Shi, D. Chen, Z. Zhong, S. Zhong, L. Zhang, S. Chen, S. J. Morrison, R. L. Maas, K. Zhang and Y. Liu, *Nature*, 2016, **531**, 323–358.
91. J. H. Park, N. Arora, H. Huo, N. Maherali, T. Ahfeldt, A. Shimamura, M. W. Lensch, C. Cowan, K. Hochedlinger and G. O. Daley, *Cell*, 2008, **134**, 877–886.
92. M. Grskovic, A. Javaherian, B. Strulovici and G. Q. Daley, *Nat. Rev. Drug Discovery*, 2011, **16510**, 915–929.
93. A. Mullard, *Nat. Rev. Drug Discovery*, 2015, **14**, 589–591.
94. M. Scudarelli, *Nature*, 2016, **534**, 310–312.
95. S. Yamanaka, *Cell Stem Cell*, 2012, **10**, 678–684.
96. D. Rajamohan, E. Matsa, S. Kalra, J. Crutchley, A. Patel, V. George and C. Denning, *Bioessays*, 2013, **35**, 281–289.

CHAPTER 8

Protein Therapeutics and Blinding Diseases

SAHAR AWWAD,[a,b] PENG T. KHAW[a,b] AND
STEVE BROCCHINI*[a,b]

[a] UCL School of Pharmacy, London; [b] National Institute (NIHR)
Biomedical Research Centre at Moorfields Eye Hospital NHS
Foundation Trust and UCL Institute of Opthalmology, London
*Email: ucnvsbr@ucl.ac.uk

8.1 INTRODUCTION

Becoming blind is hard to comprehend. It is impossible to know what it is to be blind by simply wearing a blindfold for a brief period. Most people will do everything possible to avoid going blind. Approximately 285 million people are visually impaired globally, with 90% of these people living in economically de-prived regions of the world. Blindness afflicts people from all demographics, but age-related eye diseases are becoming more common as a result of the growing elderly population.

Although we face many health issues as we age, loss of visual function dramatically and negatively affects quality of life and sense of well-being. Visual impairment is one of the most highly ranked diseases by patients. Moderate to severe disabilities in

Engineering Health: How Biotechnology Changed Medicine
Edited by Lara V. Marks
© The Royal Society of Chemistry 2018
Published by the Royal Society of Chemistry, www.rsc.org

our ageing population, which includes loss of visual function, are projected to increase by 32–54% in the UK by 2022. A person who has lost their sight must adapt to challenges imposed for every activity, especially isolation, which can be exacerbated in the elderly.

New biological medicines have been introduced for the treatment of ophthalmic disorders since the start of the 21st century. The vast majority of these medicines use monoclonal antibodies (Mabs). These are laboratory produced antibodies derived from those made by the immune system to protect the body against foreign invaders. Also known as immunoglobulins (Igs), antibodies have evolved to tightly adhere to the cells of infectious bacteria and other foreign organisms to facilitate their destruction. Over the past 30 years scientists have found ways of producing Mabs that bind specific disease targets. This has radically transformed the therapeutic landscape. Ophthalmic antibody-based medicines can now treat previously untreatable blinding age-related eye diseases and help maintain sight by slowing down the progression of chronic blinding diseases.[1] Biotechnology also offers the potential to develop new treatments to stop blinding disease progression more widely and even to regenerate damaged tissue.

8.2 THE EYE AND BLINDING DISEASE

The eye is broadly divided into two compartments called the anterior (front of the eye) and posterior (back of the eye) segments. These two eye segments are separated by the iris and lens. Light first enters the eye through the cornea and passes through the anterior chamber. The iris is the colourful part of the eye and can shrink or enlarge the pupil to regulate the amount of light entering the eye. The lens is made of protein and water and it focuses the light onto the retina. Once light goes through the lens, it then passes through the posterior segment to the retina, which is a light-sensitive tissue lining the inner surface of the eye in the posterior segment. There are two main types of specialised light-sensitive cells in the retina called rods and cones. These photoreceptors process light into biological signals that can be interpreted by the brain. Rods are sensitive to low light levels and are not able to discern colour. Cones are

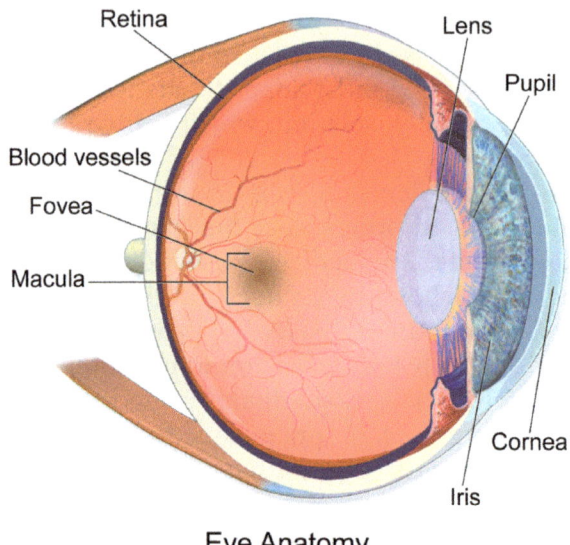

Eye Anatomy

Figure 8.1   Structure of the human eye. The anterior segment (front of the eye)
consists of the iris, cornea, pupil and lens. The posterior segment
(back of the eye) consists of the retina, vitreous humour, macula,
fovea and optic nerve. Most severe and threatening blinding diseases occur in the posterior segment.
Image by Holly Fischer, published under Creative Commons 3.0
Licence (https://creativecommons.org/licenses/by/3.0/deed.en)

more active at higher light levels and are capable of colour
vision. They provide the capacity for sharper vision than is possible with rods. The largest density of cones is located centrally
in a small region of the retina called the macula (Figure 8.1),
which provides the highest possible visual acuity. The macula
controls the central part of our field of vision, which allows us to
see details (*e.g.* to recognise faces or read). Disruption of the tissue
in the macula due to disease can cause the loss of central vision.

When light passes into our eye and interacts with a photoreceptor, a process known as phototransduction occurs. Light
causes changes to the chemical shape of pigments in the
photoreceptor that cause the activation of a protein called
transducin. This results in a sensory response that is transferred
to the retinal ganglion cells, which are neurons. Hence a light
stimulus is turned into a nerve signal, which can then be processed in the brain. Ganglion cells from the whole retina pass

into our brain *via* the optic nerve. These cells are important for vision and vision is irreversibly lost when they die. Ganglion cells are naturally lost over time, but this can be accelerated by a group of diseases called glaucoma. Death of the optic nerve also leads to permanent blindness.

8.3   COMMON BLINDING CONDITIONS

8.3.1   Cataracts and Glaucoma

Posterior segment disorders account for 75% of all ocular diseases, many of which are age-related disorders. These include cataracts, glaucoma and age-related macular degeneration (AMD). Cataracts are one of the most common causes of sight loss in the elderly. Caused by the clouding over of the lens in the eye, cataracts can be easily remedied with surgery. This is a routine surgical procedure that does not need hospitalisation. It involves the replacement of the cloudy lens with a new synthetic lens called an intraocular lens. Many intraocular lenses are made of a plastic material known as a hydrogel. Often used for contact lenses, hydrogels can be made to be soft, transparent and to hold a lot of water while also allowing the permeation of gases such as oxygen.

Unlike cataracts, glaucoma is inoperable and the main cause of irreversible blindness.[2] While glaucoma is a complex disease, the main modifiable risk factor is the intraocular pressure (IOP). Any prolonged high IOP, as can occur with glaucoma, can impair ganglion cells in the optic nerve. Thus IOP is often measured when going to the optician for an eye check-up. Most glaucoma treatment focuses on decreasing the IOP to avoid further damage to the optic nerve, as it is currently the only proven method to reduce progression of the disease.[2]

8.3.2   Age-related Macular Degeneration (AMD)

Another ophthalmic disease common among elderly patients is age-related macular degeneration (AMD), which affects approximately ten million people worldwide. The disease is most prevalent in those over 60 years of age. There are two forms of AMD: non-neovascular AMD (dry form) and neovascular AMD (wet form).

Non-neovascular AMD accounts for 85–90% of all cases of AMD. It is linked to the formation of drusen, a yellow–white extracellular material that affects the retinal pigment epithelium (RPE). Drusen can be either hard or soft. The hard drusen resemble pinpoint yellow–white lesions. The soft drusen are bigger in size and have indistinct edges and can be present in a large number in the central macula.[3] While drusen do not usually cause vision loss, their accumulation can lead to the slow and progressive loss of vision and a predisposition to AMD.

Approximately 10–15% of all AMD that cause blindness are the result of wet or neovascular AMD. This condition is linked to angiogenesis, a process where new blood vessels develop from existing blood vessels. Angiogenesis is very important for development in the womb and subsequent growth, but when it goes out of control it can cause many medical conditions, including cancer and inflammation. Wet AMD is characterised by uncontrolled angiogenesis in the macula region of the retina that is responsible for sharp vision and the ability to see detail. Blood vessels grow from under the retina (choroid) into the subretinal space. This can lead to the leakage of blood/serum into the macula, which causes tissue damage and a loss of vision. The growth of new blood vessels is often accompanied by local tissue inflammation and scarring (fibrosis), leading to severe visual impairment. Inflammatory responses result in more complicated problems. For example, they can induce abnormal blood vessels to leak, resulting in the thickening and swelling of the retina. Unwanted blood vessel growth can also result in retinal detachment.

Prior to the development of biological medicines, ocular disease tended to be treated with antioxidant and photodynamic therapy. Antioxidant therapy consists of the use of natural vitamins and other nutritional supplements. Studies have shown that vitamin C and E, beta-carotene and zinc oxide can help reduce the progression of AMD. Vitamins, however, have not been demonstrated to be useful in the treatment of wet and dry AMD.[4] Photodynamic therapy was developed in the 1990s. It involved the injection of a light-sensitive medicine, such as verteprofin, into a vein in the arm of the patient, which attaches itself to abnormal blood vessels in the macula. Following this a low-powered laser was shone into the damaged eye to activate the drug, which then destroys the abnormal vessels in the macula. The therapy was largely used to treat AMD, but was largely phased out as a result of the arrival of ranibizumab, a Mab drug, which was found to be superior in terms of safety and improvement of vision for the condition.[5]

8.3.3   Diabetic Macular Edema (DME)

Blinding disorders do not just afflict the old. One of the many causes of irreversible blindness in younger adults is diabetic macular edema (DME), which is a complication of diabetes. DME affects almost 29% of diabetic patients. The condition is typically characterised by an increase in retinal thickening in the macular area. There are two types of DME: focal and diffuse macular edema. Focal macular edema is a result of the breakdown of the blood–retinal barrier (BRB). The BRB not only guards against foreign substances entering the retinal tissues, it also controls fluid movement between the ocular blood vessels. Any break in the barrier can result in an increased leakage of intra-retinal fluid from the abnormal retinal blood vessels and capillaries. This can cause angiogenesis and inflammation, which increase vascular permeability in diabetic eyes. Diffuse macular edema involves the dilation of the retinal capillaries and swelling and damage of the retinal tissue.

8.3.4   Uveitis

Another common inflammatory eye disorder that can occur throughout the eye is uveitis. It can be caused by an autoimmune response, which is when the immune system mistakenly attacks healthy tissues. Uveitis takes different forms. Autoimmune uveitis involves the inflammation of the uveal tract that includes the iris and the connective tissue in the retina called the choroid. Infectious uveitis accounts for 30–50% of all cases of uveitis in developing countries. It is caused by infections with the herpes virus and toxoplasmosis and, to a lesser extent, tuberculosis and syphilis. Toxoplasmosis results in 40–50% of children getting posterior uveitis.

8.4   BIOLOGICAL TREATMENTS

A number of new biological drugs have now been developed for the treatment of ophthalmic conditions. Many of these target the vascular endothelial growth factor (VEGF). VEGF is a signalling protein that helps restore the supply of oxygen to tissues when blood circulation is inadequate and is implicated in angiogenesis. Inhibiting VEGF has been clinically shown to slow the progression of medical conditions driven by angiogenesis.

Pegaptanib (Macugen®) was the first anti-angiogenic therapy, approved in 2004 for the treatment of neovascular AMD. The approval of pegaptanib was a milestone in drug development as it was the first aptamer to be successfully developed as a therapeutic agent for humans. Discovered in the early 1990s, aptamers are a specific class of nucleic acid molecule. Aptamers have the advantage of not only being small in size but of also being highly specific in their target and not provoking immune responses.

Since the introduction of pegaptanib, there have been a number of newer generation anti-VEGF therapies for the treatment of wet AMD. This includes ranibizumab [fragment antigen binding (Fab)] and bevacizumab (full Mab), and aflibercept (recombinant fusion protein). All three drugs have revolutionised the treatment of AMD.[6] The advantage of the Mab drugs is that they target the growth factors and mediators that drive inflammation and angiogenesis that underlie such disorders. Many of these drugs have been clinically approved for treating inflammation in other parts of the body. In addition to helping with AMD, the Mab drugs are used to treat DME and posterior uveitis, which was previously treated with steroids and immunosuppressive agents.

8.5   ADMINISTERING MAB THERAPEUTICS TO THE EYE

8.5.1   Challenges With Current Routes of Administration

Many common blinding conditions necessitate treatment that can reach the back of the eye. This is a major challenge. Eye drops are the most common and preferred treatment route for ocular diseases. They are widely used to treat conditions on the outside of the eye, such as dry eye, conjunctivitis and corneal infections, which are often caused by wearing contact lenses. Eye drops are not a viable route for administering antibody-based drugs to the back of the eye. Several factors, such as tear film, tear drainage and blinking, limit the volume of eye drops that can be administered. Many physical and biochemical barriers within the eye also prevent the drug reaching the back of the eye, making it difficult to achieve reproducible and high drug doses.

Most Mab treatments are administered by injections into the bloodstream. Drugs given orally are usually degraded in the gastrointestinal tract (GIT). This is because they are treated like food proteins and so are broken down by acid and enzymes. Even if a Mab could be formulated to be protected from the harsh environment of the GIT, it could not pass through the cells linings of the GIT into the bloodstream, mainly due to its large size. The majority of an injected therapeutic protein remains in the blood circulation system until it is cleared by the kidneys (renal clearance) or metabolised (*i.e.* degraded).

Medicines that are administered by injection or orally are distributed throughout the body. These are known as systemically administered drugs. However, entry from the blood system into the eye is limited because much like with the brain, a systemically distributed drug must overcome the BRB to enter the eye. Outer and inner BRBs also hinder the flow of drugs into the retina. Very high systemic doses are thus required to achieve therapeutic concentrations within the eye. Systemic drug dosing results in the drug being distributed throughout the body, causing unwanted side effects. Often one eye requires treatment, so systemic drug dosing results in both eyes being dosed with the drug, which risks potential side effects in the healthy eye.

8.5.2 Direct Intravitreal (IVT) Injections to the Vitreous Humour

The penetration of therapeutic proteins, like Mabs, into the eye is obstructed by the structure and biophysiological properties of the cornea. As with the rest of the eyeball, the cornea is not permeable to the outside environment, *e.g.* dust and bacteria. This is important because the inside of the eye is 'privileged' and must remain sterile because the eye is highly vulnerable to infection. Some medicines are able to permeate the cell membranes of the cornea to some degree, but not biotechnology-derived therapeutic proteins like Mabs, which are water-soluble molecules. The five or six layers of cells on the front of the cornea are tightly packed epithelial cells that are resistant to the permeation of water-soluble molecules. The cornea is transparent so that light can enter the eye and be projected onto the retina. There are no blood vessels in the cornea to block the incoming light. To nourish the cornea, aqueous fluid is secreted from the ciliary body $(2.0–2.5\ \mu l\,min^{-1})$ from behind the lens into the vitreous humour. The vitreous humour (Figure 8.2) is a viscous and transparent gel that it is important for the maintenance and metabolism tissues in the back of the eye. The vitreous humour consists mostly of water (98–99%) and other components. It helps protect the eye from trauma, inhibits angiogenesis and

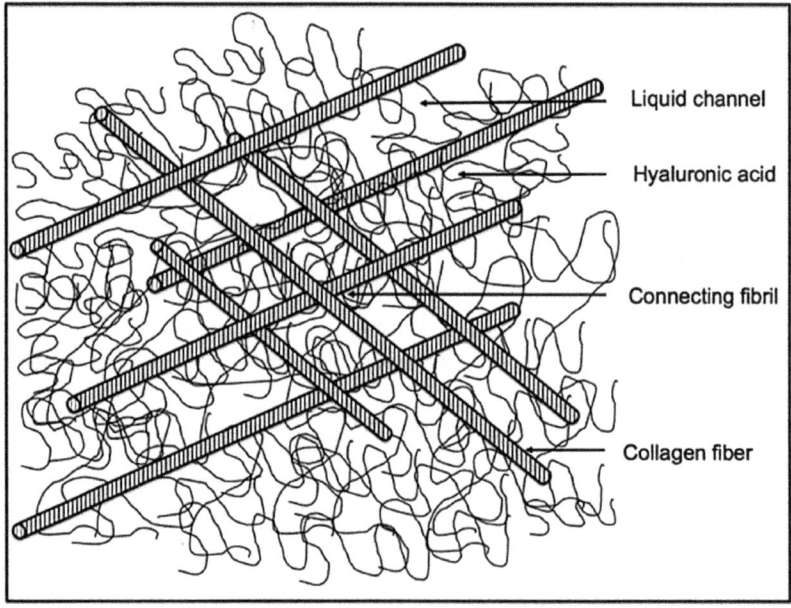

Figure 8.2   Schematic representation of human vitreous humour. The vitreous humour is made of packed collagen fibril bundles and the terfibrillar spaces are filled with hyaluronic acid and water molecules. The vitreous humour has become a popular site for IVT injections that enable a therapeutic drug dose to reach the back of the eye.

coordinates eye growth. The vitreous humour also regulates eye growth and the shape of the eye during development, including the maintenance of the transparency needed for sight.

The liquid in the front of the eye in the anterior chamber is much less viscous than the vitreous humour. It is essentially comprised of aqueous humour secreted from the ciliary body, which flows into the anterior chamber to nourish the cornea and the front of the eye. The aqueous humour then predominantly exits the eye at the edges of the cornea through what is known as the trabecular meshwork. Mass exchange within the eye is therefore dominated by aqueous flow. If a medicine does get past the cornea window when using an eye drop, it then needs to diffuse upstream against the aqueous outflow in the anterior chamber to then get into the posterior segment. Most protein drugs that might permeate the cornea would simply exit the eye *via* the aqueous outflow. The lack of corneal permeation and aqueous outflow make it very difficult to use eye drops to achieve a reproducible and sufficiently high dose of a protein therapeutic in the back of the eye.

In order to achieve a reproducible and sufficient dose in the back of the eye, medicines are currently injected directly into vitreous liquid in the back of the eye. This is the key common method used for the delivery of therapeutic proteins. Direct injections known as intravitreal (IVT) injections are routinely conducted in ophthalmology clinics. Good practice is very important to avoid the risk of infection and other complications that are possible with IVT injection. The injection volume is usually 50 µl (0.05 ml). Direct IVT injection avoids all blood–ocular barriers and ensures that an optimal and reproducible dose can be achieved in the back of the eye.

Proteins diffuse much more slowly in the viscous vitreous liquid in the posterior chamber than low-molecular-weight molecules after IVT injection. Most low-molecular-weight molecules that are injected into the eye will permeate the retina, so these molecules generally clear by both the aqueous outflow and retina choroid sclera pathways. Since therapeutic proteins diffuse much more slowly in the vitreous humour and clear predominantly by aqueous outflow, they last much longer in the back of the eye, often days, than low-molecular-weight drugs, which persist for a matter of hours. Repeated IVT injections are required to treat chronic blinding conditions, so minimising the frequency of IVT injections is very important to avoid possible complications and reduce inconvenience to the patient and healthcare costs. Efforts are now underway to improve the properties of Mabs to reduce the frequency of IVT injections as well as find alternative means of administering them.

8.6   THE CHANGING FACE OF MAB DRUGS

Over the last twenty years different fragments from Mab molecules have been used for treatment of various diseases. These fragments are aimed at resolving the poor penetration of Mabs across cell membranes to increase their efficacy and reduce the doses needed. Two key regions from the antibody are used in treatments. The first is taken from the portion of the antibody that specifically binds to a receptor on the cell of foreign substance, known as an antigen. The abbreviation for this region is Fab, which stands for fragment antigen-binding. Each antibody has two Fabs (see Figure 8.3). They form the two tips at the top of the Y-shape of the molecule. They are linked by another region, the stem of the Y. This is called the fragment crystallisable region, or Fc for short. One of the functions of the Fc region is to activate the host immune mechanisms and complement-dependent

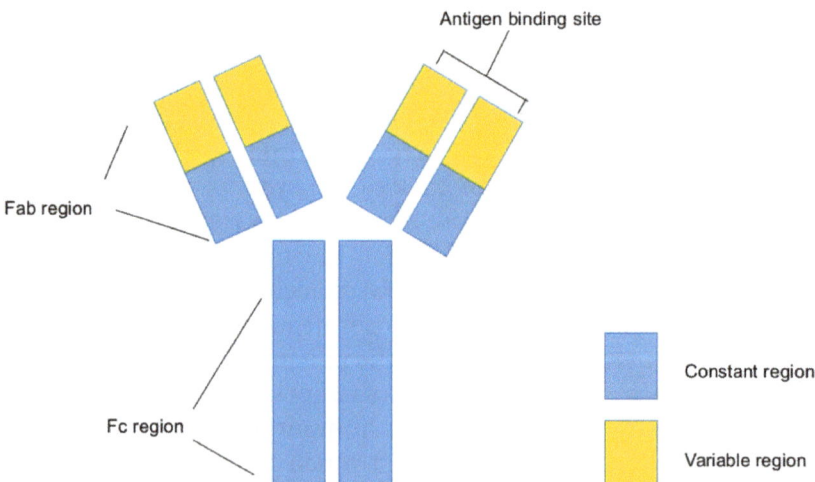

Figure 8.3   Structure of an antibody showing Fab and Fc regions.

cytotoxicity. These are effector functions that serve to destroy and eliminate a pathogen. It can also stimulate the secretion of cytokines, small proteins that aid cell communication and facilitate the movement of cells towards sites of infections, inflammation and trauma to aid recovery. The Fc also helps recycle circulation of the antibody in the blood.[7]

One example of a Fab drug is ranibizumab (Lucentis®, Roche, Genentech), which was approved by the US Food and Drug Administration (FDA) in 2006. It was specifically developed to be injected into the vitreous humour of the eye to treat AMD.[5,8] The rationale for developing a Fab rather than a full Mab was that the smaller size of the Fab would allow better binding to VEGF in the retina. The development and use of ranibizumab has revolutionised the treatment of wet AMD and other neovascular conditions that often resulted in blindness or considerable loss of visual function (*e.g.* diabetic retinopathy and retinal vein occlusion). Different therapeutics have also been developed for different medical conditions using the Fc region of the antibody. The Fc region of the antibody is generally combined with another peptide or protein so that it can bind to the therapeutic target of interest. Such drugs are known as Fc-fusion proteins. Fc-fusion proteins are made to provide a means to increase the absorption, circulation, distribution, metabolism and excretion of the drug within the body.

So far several Fc-fusion proteins have entered the market since the first one was approved in 1998.[9] In 2012 the FDA approved ziv-aflibercept (Zaltrap®, Sanofi) for treating metastatic colorectal cancer that has become resistant to first-line treatment. A different formulation of ziv-aflibercept called aflibercept (Eylea®/VEGF Trap-Eye, Regeneron) is now used to inhibit ocular angiogenesis to treat AMD and DME. The drug targets human VEGF-A and B receptors and another growth factor called placental growth factor (PGF). In addition to the use of Mabs and their fragments, bispecific antibodies are being explored as a form of treatment. In contrast to conventional Mabs, which bind to only one specific receptor on an antigen, bispecific antibodies bind to two different antigens or two different receptors on the same antigen. Bispecific antibodies have several potential advantages: (i) they can redirect specific immune cells to the tumour cells to enhance tumour killing, (ii) they can enable simultaneous blocking of two different mediators/pathways that cause a disease and (iii) they can potentially increase binding specificity by interacting with two different cell-surface antigens instead of one.

8.7 STEM CELLS

Advances in ophthalmic medicine are not only reliant on the use of Mabs. In recent years stem cells have also been explored as a potential avenue for treatment. Explored in detail in Chapter 7, stem cells comprise some of the body's master cells, which have the ability to grow into many different cell types. Stem cells are important to the renewal and repair of tissue in the body. Such cells are found in a variety of sources, including the bone marrow, blood from the umbilical cord and embryos. Recent progress in stem cell research is generating particular optimism for the treatment of dry AMD. At present there is no treatment that can reverse dry AMD, though some vitamins and anti-oxidants can slow progression.[10] Stem cells, however, could help treat severe retinal degeneration involved in the condition, thereby restoring sight.[11] The RPE is the major target of AMD. Attempts have therefore been made to replace or graft a new RPE at the macula. Different sources of RPE cells have been used in these procedures. In one study, RPE cells derived from embryonic stem cells were reported to increase to likelihood of photoreceptor and central visual rescue.[12] Several trials are currently underway to treat diseases like AMD and retinitis pigmentosa using human embryonic, foetal and umbilical cord tissue-derived stem cells and bone marrow-derived stem cells.[10]

Stem cell transplantation for retinal diseases has made a lot of progress in the last decade to phase I/II clinical trials. Many issues, however, still need to be resolved before stem cells can be applied in the clinic. Some of these are the biological risks and technical difficulties associated with differentiation culture procedures.[11] More understanding of retinal regeneration phenomenon will help provide more information on the cellular basis of retinal regeneration and expand the possibilities for cell therapies for retinal degenerative diseases.[11]

8.8   CILIARY NEUTROPHIC FACTOR

In addition to stem cells, researchers are exploring the use of ciliary neutrophic factor (CNTF) for eye treatment. CNTF is a protein that acts as a signal in times of stress and injury to protect neural tissues. It helps protect photoreceptors in times of trauma within the retina and it is important in slowing down vision loss due to photoreceptor death. A company called Neurotech Pharmaceuticals is currently developing an implantable platform with human cells genetically engineered to secrete therapeutic doses of CNTF into the back of an eye to help treat retinal degenerative diseases. The implant is called NT-501 and results from clinical trials have indicated that the platform can help reduce photoreceptor degradation in patients with retinitis pigmentosa.[13–15]

8.9   THE CHALLENGES BEHIND THE CLINICAL DEVELOPMENT OF OPHTHALMIC DRUGS

8.9.1   Anatomical Differences Between Human and Animal Models

Before any new medicine can be used in humans it needs to first undergo a series of tests to assess its biological action, metabolism and toxicity. This is usually done through animal testing, known as *in vivo* testing. *In vivo* is defined as processes that take place in a living organism. Many anatomical differences between human and animal eyes have to be carefully considered. Table 8.1 lists the differences in vitreous volume and aqueous outflow in

Table 8.1   Anatomical and aqueous flow differences between human and animal models.

Species	Vitreous volume (ml)	Aqueous flow ($l\,min^{-1}$)
Human	4.0	2.5–3.0 2.8 (20–30 years) 2.4 (>60 years)
Cat	2.4–2.7	5.0–5.9
Dog	1.7	5.2
Horse	26.15	Not applicable
Monkey	3.0–4.0	2.8
Mouse	5.6×10^{-5}	0.18
Pig	3.5	3.7
Rabbit	1.0–1.5	2.0
Rat	$1.3–5.4\times10^{-2}$	0.35

the most commonly used animal models. Rabbits are a species of choice for most evaluation of drugs for eye diseases because their eyes have similar anatomical features to human eyes.[16]

8.9.2 Overcoming Anti-drug Antibodies (ADAs) in Animals

Testing drugs and formulations in animals is not straight-forward. Notably an immune response can develop, called anti-drug antibodies (ADAs), to therapeutic proteins that have been designed for use in humans. ADAs lead to rapid clearance of the candidate protein and can also cause acute hypersensitivity or infusion reactions. All of this can hinder accurate evaluation and optimisation of a therapeutic protein drug in animals. The limitations imposed by ADAs are potentially more problematic for longer-acting therapeutic proteins, which would be expected to be more susceptible to an immunogenic reaction over longer exposure periods. In the light of ADAs, animal models may only be appropriate for characterising new biotechnology products in carefully designed studies that only utilise candidate products that have already gone through a significant degree of development.

8.9.3 Difficulty in Stability Studies

One of the criteria measured in testing is the stability of a therapeutic protein. This can be very difficult to determine using small samples obtained from animals. Many therapeutic proteins easily aggregate and misfold over time, whether in storage or during release from a long-acting formulations *in vivo*. Isolation and purification processes that are required to obtain the therapeutic protein from biological fluids must be validated to ensure that the stability of the therapeutic protein can be determined. Although curve-fitting software can be developed to use animal data to predict or estimate human clearance times, the critical issues related to ADA formation and the lack of practical methods to evaluate protein stability need careful consideration.

8.9.4 *In Vitro* Testing Challenges

Many new medicines and formulations can now be tested in non-living organisms, also known as *in vitro* techniques/models.

In vitro models have been extensively used in drug development. They help maintain standards in manufactured products by testing a sample against established specifications. This is also known as quality control (QC). Drug property data obtained from *in vitro* experiments have been used to predict *in vivo* responses. Many types of *in vitro* models are also used in drug development. Until recently, however, there have not been any *in vitro* models that can be used to estimate the human clearance time of therapeutic proteins from the back of the eye. In most cases a test tube was used to evaluate the static partitioning of a candidate drug between water and oil. This is inadequate for studying the context of aqueous outflow and predicting the clearance time needed for the development of therapeutic proteins for ophthalmic use.

8.9.4.1 Current In Vitro *Models.* The advent of protein therapeutics has resulted in the need for new *in vitro* models to predict human clearance times. Until now an apparatus called an USP-4 has been used to evaluate poorly soluble drugs and extended-release tablets.[17] However, the volume, flow rate and geometry of the USP apparatus do not provide sufficiently accurate representations of the human eye. In addition to the USP-4, a single-compartment, non-flow model for eye-movements (saccades), is used to evaluate other aspects of drug distribution in the vitreous humour. This is important because eye movements can influence drug distribution in the vitreous humour.[18]

A great deal of research is now being carried out to develop new strategies to increase the vitreous residence times of protein therapeutics to avoid the use of high-frequency IVT injections. One team at University College London have recently created a new *in vitro* model to help in the pre-clinical development of new ocular drugs and their delivery mechanisms. The PK-Eye is a two-compartment, aqueous outflow model scaled to the human eye.[19] It has been specifically designed to mimic the intraocular aqueous outflow to estimate the clearance of therapeutic protein formulations from the vitreous cavity (Figure 8.4).

The PK-Eye is currently being used to study the stability of proteins in order to develop strategies to maintain the presence of the protein in the posterior cavity for longer periods, for

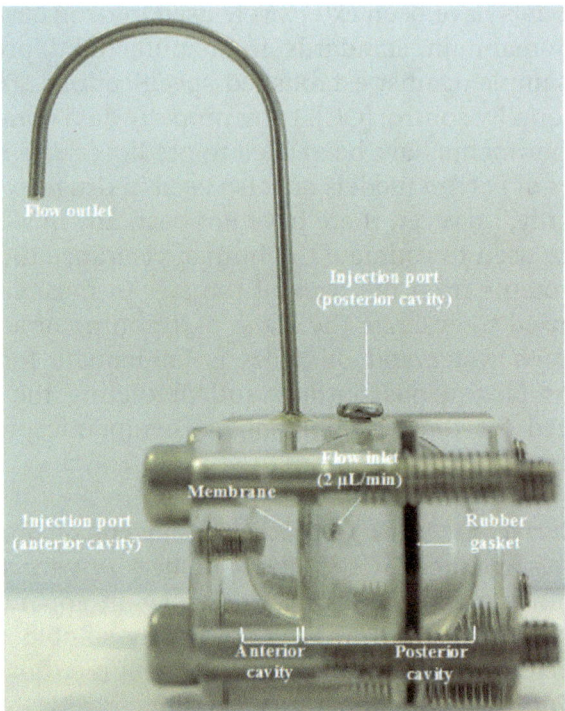

Figure 8.4 Design of the PK-Eye model used for the evaluation of protein
therapeutics and low-molecular-weight drugs. The model com-
prises of two cavities: anterior (0.2 ml) and posterior (4.2 ml) separated by a membrane. The inlet port allows a flow of 2.0 μl min^{-1}
of PBS, pH 7.4 with 0.02% sodium azide. A rubber washer is fitted
within the model to prevent leakage. Two injection ports are present
in both cavities to allow drug administration. An outlet port allows
sample collection for further analysis.
Reprinted from *Journal of Pharmaceutical Sciences*, Volume 104,
S. Awwad, A. Lockwood, S. Brocchini, and P. T. Khaw, "The PK-Eye:
A Novel *In Vitro* Ocular Flow Model for Use in Preclinical Drug
Development," 3330–3342, Copyright 2015, with permission from
Elsevier.[19]

example in the form of implants and sustained-release formulations. One of its advantages is that it makes it easier to evaluate
protein function and stability. This is difficult to accomplish
using animal models and is especially important for dosage
forms that have longer clearance times that can extend over a
2–3-month period. The PK-Eye model has many of the features
needed to become a practical *in vitro* model with the capacity to

contribute to research efforts focused on the development of new protein therapeutics.[19] The PK-Eye has also been used to study formulations containing low-molecular-weight drugs indicated for the treatment of ocular diseases and infections.[20]

8.10   COST ISSUES

One of the key issues in the development of drugs is how much can be charged for the end product. In recent years two Mab drugs, ranibizumab (Lucentis®) and bevacizumab (Avastin®) have hit the headlines because of their contrasting price levels. Both drugs are used as treatments for wet AMD. Out of the two drugs, ranibizumab is more expensive. A typical dose of ranibizumab costs $2000 a dose, whereas bevacizumab costs $50.[21] The lower cost of bevacizumab makes it potentially more attractive than ranibizumab, but it has never been specifically licensed to treat wet AMD. It was approved in 2004 to treat cancers, such as metastatic colorectal, non-small-cell lung cancer, renal cell cancer and glioblastoma. The FDA or European Medicines Agency (EMA) approves a medicine for specific indications depending on the results of clinical trials. The drug cannot be promoted for other uses without appropriate clinical trials and licencing. Clinicians are nonetheless permitted to use them 'off-label' in different circumstances although the legal framework for this varies in different countries. Ophthalmologists started to use bevacizumab off-label around 2005 while awaiting FDA approval of ranibizumab. Evidence had already demonstrated the drug not to be toxic to the eye and to provide some clinical benefit in the treatment of neovascular AMD.[22,23]

One of the problems in using bevacizumab for eye disorders is that it has not been specially formulated for such treatment. Bevacizumab is presented in a vial containing 400 mg of antibody at a concentration of 25 mg ml^{-1}. In order to be administered into the eye it needs to be transferred under aseptic conditions into ready-to-use 1.0 ml plastic syringes for IVT injection by compounding pharmacies for local distribution. Pharmacy compounding constitutes the preparation of personalised medications for patients. In this context individual ingredients are mixed together in the exact strength and dosage form required by the patient. The dose of bevacizumab (1.25 mg) for IVT injection was established simply by using 50 μl of the solution from the vial.

Local compounding carries a number of risks. Importantly, there can be a wide variation in how bevacizumab is transferred to the syringes from the vial. This can have a knock-on effect in terms of dosage preparation and can result in contamination and protein aggregation, which can cause adverse reactions *in vivo*. To avoid these risks, there have been reports of 'multiple use' from a vial of bevacizumab to treat patients consecutively. However, there is the risk of infection if the vial is punctured multiple times. Indeed, it has been connected with an increased incidence of endophthalmitis, an inflammation of the internal coats of the eye. Furthermore, differing lubricants used in various pre-packed syringes have been reported to cause inflammation.

The preparation of bevacizumab for administration to patients still remains a problem today. Nonetheless the high cost of ranibizumab means that the IVT administration of bevacizumab will continue in many parts of the world, especially in resource-limited regions and especially for older patients whose overall health and social care costs are already high and are expected to increase.

While the costs for using bevacizumab are much less than those for ranibizumab, the National Institute for Health and Care Excellence considers the compounding of bevacizumab into syringes followed by storage prior to ophthalmic use to be unlicensed use of bevacizumab. Both ranibizumab and bevacizumab were developed by Roche, and while the FDA has indicated that bevacizumab can be approved for IVT injection to treat blinding neovascular disease, the company has, to date, been unwilling to seek this approval for bevacizumab.

8.11 CONCLUSION

According to the World Health Organisation (WHO), AMD is the leading cause of blindness in developed countries and ranks third after cataract and glaucoma globally.[24] More than a decade ago, treatment options were limited, with laser photocoagulation and verteporfin photodynamic therapy being used extensively to treat neovascular AMD. However, neither of these two is effective for all patients with neovascular AMD nor does it improve visual acuity.[25] Better understanding of the underlying pathophysiology of ocular diseases has led to the development of more effective treatments.[26]

The understanding of AMD and its pathogenesis has changed over decades of investigation.[24] Treatments inhibiting VEGF have been proven to stop loss of vision in more than 90% of patients, with vision improvement in one third. Anti-VEGF medicines for treatment of intraocular neovascularization have shown improvements in signs of disease and quality of life.[1] Therapeutic biologics registered for ophthalmic use by IVT injection comprise a PEGylated-aptamer (pegaptanib), an antibody fragment (ranibizumab) and an Fc fusion (aflibercept). The full length Mab, bevacizumab is also widely used as an unlicensed medicine to treat AMD. Despite the remarkable advances made in anti-VEGF therapies, most patients require an increased frequency of re-injections and regular long-term follow-up. Extensive efforts have been made to develop advanced drug delivery devices to reduce treatment burdens and to improve patient compliance.[27]

It is anticipated that ophthalmic protein-based medicines, which tend to be potent and have a rapid onset of action, will continue to be developed as the molecular mechanisms involved in blinding diseases become better understood. The dosing, frequency and threshold for treatment and re-treatment are leading to the identification of novel therapeutics with longer duration of action and higher efficacy. New pharmacological agents have been developed to target various areas of the VEGF pathway.[6] Though challenges remain, the future of treating AMD appears promising. A number of clinical questions still remain regarding the treatment of AMD, despite the success of these biological medicines. The dosing, frequency and threshold for treatment and re-treatment are leading to the identification of novel therapeutics with longer duration of action and higher efficacy. Recently the ophthalmic community has provided guidelines on the treatment of AMD to help clinicians prevent over- and under-treatment with anti-VEGF medicines.[27]

ACKNOWLEDGEMENTS

We are grateful for funding from the NIHR Biomedical Research Centre at Moorfields Eye Hospital NHS Foundation Trust and UCL Institute of Opthalmology, Moorfields Special Trustees, the Helen Hamlyn Trust (in memory of Paul Hamlyn), UK Medical Research Council, Fight for Sight and the Michael and Ilse Katz Foundation.

REFERENCES

1. E. B. Rodrigues, M. E. Farah, M. Maia, F. M. Penha, C. Regatieri, G. B. Melo, M. M. Pinheiro and C. R. Zanetti, *Prog. Retinal Eye Res.*, 2009, **28**(2), 117–144.
2. P. Bettin and F. Di Matteo, *Ophthalmic Res.*, 2013, **50**(4), 197–208.
3. J. Ambati, B. K. Ambati, S. H. Yoo, S. Lanchulev and A. P. Adamis, *Surv. Ophthalmol.*, 2003, **48**(3), 257–293.
4. N. M. Bressler, S. B. Bressler, N. G. Congdon, F. L. Ferris, D. S. Friedman, R. Klein, A. S. Lindblad, R. C. Milton and J. M. Seddon, *Arch. Ophthalmol. (Chicago, IL, U. S. 1960)*, 2003, **121**(11), 1621–1624.
5. N. Ferrara, L. Damico, N. Shams, H. Lowman and R. Kim, *Retina*, 2006, **26**(8), 859–870.
6. P. S. Prasad, S. D. Schwartz and J. P. Hubschman, *Maturitas*, 2010, **66**(1), 46–50.
7. S. Mitragotri, P. A. Burke and R. Langer, *Nat. Rev. Drug Discovery*, 2014, **13**(9), 655–672.
8. P. J. Rosenfeld, D. M. Brown, J. S. Heier, D. S. Boyer, P. K. Kaiser, C. Y. Chung and R. Y. Kim, *N. Engl. J. Med.*, 2006, **355**(14), 1419–1431.
9. W. R. Strohl, *BioDrugs*, 2015, **29**(4), 215–239.
10. C. M. Ramsden, M. B. Powner, A. J. F. Carr, M. J. K. Smart, L. da Cruz and P. J. Coffey, *Development*, 2013, **140**, 2576–2585.
11. S. Jeon and I. H. Oh, *BMB Rep.*, 2015, **48**(4), 193–199.
12. S. D. Schwartz, J. Hubschman, G. Heilwell, V. Franco-Cardenas, C. K. Pan, R. M. Ostrick, E. Mickunas, R. Gay, I. Klimanskaya and R. Lanza, *Lancet*, 2012, **379**(9817), 713–720.
13. T. Thrimawithana, S. Young and C. Bunt, *Drug Discovery Today*, 2011, **16**(5–6), 270–277.

14. P. A. Sieving, R. C. Caruso, W. Tao, H. R. Coleman, D. J. S. Thompson, K. R. Fullmer and R. A. Bush, *Proc. Natl. Acad. Sci. U. S. A.*, 2006, **103**(10), 3896–3901.
15. J. F. Girmens, J. A. Sahel and K. Marozova, *Intractable Rare Dis. Res.*, 2012, **1**(3), 103–114.
16. K. D. Rittenhouse and G. M. Pollack, *Adv. Drug Delivery Rev.*, 2000, **45**(2–3), 229–241.
17. D. Browne and S. Kieselmann, *Dissolution Technol.*, 2010, **17**, 12–14.
18. C. Loch, S. Nagel, R. Guthoff, A. Seidlitz and W. Weitschies, *Biomed Technol.*, 2012, **57**(1), 281–284.
19. S. Awwad, A. Lockwood, S. Brocchini and P. T. Khaw, *J. Pharm. Sci.*, 2015, **104**(10), 3330–3342.
20. A. Baskakova, S. Awwad, J. Q. Jiménez, H. Gill, O. Novikov, P. T. Khaw, S. Brocchini, E. Zhilyakova and G. R. Williams, *Int. J. Pharm.*, 2016, **502**(1–2), 208–218.
21. P. Whoriskey and D. Keating, 'An effective eye drug is available for $50. But many doctors choose a $2, 000 alternative,' *The Washington Post*, 7 Dec 2013.
22. U. Chakravarthy, S. P. Harding, C. A. Rogers, S. M. Downes, A. J. Lotery, S. Wordsworth and B. C. Reeves, *Ophthalmology*, 2012, **119**(7), 1399–1411.
23. D. Martin and M. Maguire, *N. Engl. J. Med.*, 2011, **364**(20), 1897–1908.
24. J. W. Miller, *Am. J. Ophthalmol.*, 2013, **155**, 1.
25. P. J. Rosenfeld, S. D. Schwartz, M. S. Blumenkranz, J. W. Miller, J. A. Haller, J. D. Reimann, W. L. Greene and N. Shams, *Ophthalmology*, 2005, **112**, 1048.
26. M. W. Stewart, *Expert Rev. Clin. Pharmacol.*, 2014, **7**, 167.
27. M. Amadio, S. Govoni and A. Pascale, *Pharmacol. Res.*, 2016, **103**, 253.

CHAPTER 9

Synthetic Biology: A Game Changer?

PAUL RACE

University of Bristol, UK
Email: paul.race@bristol.ac.uk

9.1 INTRODUCTION

The chemical sciences went through a revolution in the mid 19th century. Researchers began to turn their attention away from the isolation and characterisation of natural molecules towards the feasibility of synthesising chemical compounds, including those not found in nature, from their constituent chemical building blocks. Suddenly life-saving medicines, agrochemicals and a plethora of other useful molecules could be produced with relative ease and in significant quantities. This paradigm shift established the foundations of a field now known as synthetic chemistry, a research discipline that has indelibly changed our lives and how we live them forever.

Now, over 150 years later, the emerging science of synthetic biology, or SynBio for short, promises another scientific revolution of similar proportions. In contrast to synthetic chemistry, however, synthetic biology aims to exploit the very fabric of

Engineering Health: How Biotechnology Changed Medicine
Edited by Lara V. Marks
© The Royal Society of Chemistry 2018
Published by the Royal Society of Chemistry, www.rsc.org

life, DNA, to deliver game-changing solutions to our most pressing global problems. It has the potential to enhance human health through the development of diagnostic devices or novel medicines, to produce food to feed our growing global population, and even to provide 'green' fuels to power our everyday lives. As momentum grows, there is an increasing belief that this burgeoning field will be instrumental in tackling our greatest societal challenges.

Despite this, synthetic biology is not without its detractors. Some consider the notion of engineering biology to have insidious overtones and call into question the motives of its practitioners. Therefore, for reasons both scientific and ethical, synthetic biology represents a fascinating topic for investigation. In this chapter I provide an introduction to synthetic biology, exploring its origins and charting its emergence and evolution. I outline how synthetic biology differs from more traditional biotechnologies, debunking fact from fiction and in doing so highlight a number of key technological breakthroughs that are making the predictable engineering of biology a realistic goal. I will touch on the broader implications of synthetic biology, the ethical questions that the approach raises, and how the new science of SynBio is beginning to redefine our approach to drug discovery and healthcare. Finally, I will cast an eye to the future, imagining where synthetic biology might take us, and in doing so identify obstacles that will need to be overcome if this new field of scientific research is ever to truly deliver.

9.2 WHAT'S IN A NAME? THE ORIGINS OF SYNTHETIC BIOLOGY

Although the term 'synthetic biology' has only entered common usage within the past decade, it was in fact first coined in 1912 by the French biophysicist Stéphane Leduc.[1] Leduc believed that some forms of life could, in principle at least, be 'created' in a laboratory, using a combination of chemical and physical processes. Leduc saw this approach as a natural extension of synthetic chemistry, noting "Why is it less acceptable to seek how to make a cell than how to make a molecule?". By synthesising living matter, Leduc proposed one might not only gain a comprehensive understanding of the fundamental principles of biology, but also develop the means to create 'non-natural' living systems that could be of use to human kind. He described comprehensive methods for the generation of living cells from inanimate materials, and suggested how such cells could be formed so as to retain the shape, size and functions of their natural counterparts. These synthetic approaches, he suggested, did not need to be restricted to simple cells, but could, potentially, be used to construct tissues or more complex biological structures. At the time Leduc's ideas were considered highly controversial and attracted damming criticism by his fellow scientists. The notion of 'synthetic' life was considered fanciful at best, and the technical capabilities required to engineer even the simplest biological systems were assumed to be many decades away.

Much more comprehensive understanding of biology was needed before Ledruc's vision could become a reality. One scientific discovery in particular was to prove key to realising Leduc's dream of 'synthetic biology': Watson and Crick's uncovering of the structure of DNA in 1953.[2] This breakthrough provided the first glimpse into the blueprint of life, revealing how the elegant double helix structure, composed of a series of repeating paired building blocks called adenine (A), thymine (T), guanine (G) and cytosine (C) nucleotides, formed the basis of a duplicable, translatable, genetic code. Watson and Crick's work prompted a stepchange in our understanding of biology, ushering in a new era of scientific breakthroughs and technical advances.

The significance of DNA's distinctive chemical architecture was revealed soon after, when Crick, Brenner, Barnett and Watts-Tobin showed that DNA nucleotides function in triplicate, in units known as codons, with trios of neighbouring As, Ts, Gs and Cs, encoding individual amino acid molecules.[3] This discovery revealed the link between the genetic information contained within DNA and the structures and functions of proteins, the key components of all living cells. Although the triplet nature of the genetic code was initially demonstrated in bacteriophages, simple viruses that infect single-cell microorganisms, it was subsequently shown to be universally conserved in all forms of life on Earth. This finding confirmed that form and function in biology is dictated at the level of DNA sequence and raised the intriguing possibility that by manipulating DNA, one could, in turn, change the structures and behaviours of living systems.

In the late 60s and early 70s scientists began to discover natural biological tools for manipulating DNA. The first of these were restriction enzymes, proteins capable of cutting DNA at specific locations, which were isolated and characterised.[4] Next came the discovery of DNA ligases, enzymes able to stitch together fragments of DNA.[5] Suddenly researchers had the ability to construct nucleotide sequences of their choosing from component DNA pieces. Soon after, methods for the sequencing of DNA were described, permitting the order of individual bases within strands of DNA to be determined.[6] As a result, scientists were now able to both read, and, in an albeit basic fashion at least, write DNA.

By the 1980s researchers had access to a vast array of chemical and biological tools that could be used to manipulate and modify DNA with relative ease. The implications of this were considerable, and the age of recombinant DNA technology and molecular biology was upon us.[7] The first 'genetically engineered' microorganism was patented, a modified *Pseudomonas* bacterium capable of decontaminating oil spills.[8] Globally, billions were invested in new spinout companies and their burgeoning technologies, and recombinant DNA became big business.[9] Over time, some succeeded but many failed. The inherent complexities of biology made its exploitation, through the manipulation of DNA, at best challenging or at worst intractable. The limitations of our understanding of biological systems were exposed and frustrations grew. There was always a nagging feeling that maybe expectations had not quite been met, and maybe, just maybe, there was a better way.

By the early 2000s molecular biology was undergoing a major transformation as a result of a number of developments. Firstly, the cost of DNA sequencing began to fall at an astonishing rate. Reading DNA was no longer a costly exercise, and it was becoming possible to sequence not only selected pieces of DNA but the genomes of whole organisms. In 2000 the cost of sequencing an entire genome was in the region of £50 million, today that same genome sequence will set you back less than £3000.[10] In 2003 advances in DNA sequencing delivered the first draft of the human genome and personalised medicine was born.[11] Secondly, the traditional tools of biotechnology, restriction enzymes and ligases, once considered essential for the assembly and manipulation of DNA, had begun to be usurped. Now DNA synthesis was beginning to reign supreme. DNA sequences could be synthesised quickly and cheaply using chemical methods without any requirement for conventional molecular biology approaches.[12] Moreover, gene synthesis methods were not limited in scope to small discrete pieces of DNA, whole genomes could be produced and 'booted up' in host cells. This was helped by the availability of computational design tools and software, which had become commonplace in research labs, thereby empowering biological scientists to use predictive design methods before even setting foot in the laboratory to begin culturing bacterial strains or performing DNA extractions. Sequences of DNA morphed from stretches of As, Ts, Gs and Cs into 'biobricks', distinct functional units that could be fused and exchanged to build control circuits, toggle switches, biosynthetic pathways and fluorescent sensors in living cells. Researchers soon had the ability to add or remove large stretches of DNA into or out of the genomes of cells through the discovery of gene editing tools.[13] In addition, liquid handling robots, that can accurately perform many hundreds of thousands of experiments in parallel, started to appear in laboratories around the world. Suddenly researchers could work more quickly and efficiently and results became more reproducible. With the experimental tools needed to design, build and test new biological systems becoming available, the era of the purposeful design and engineering of biology was upon us, the era of biotechnology 2.0, the era of synthetic biology.

9.3   ENGINEERING BIOLOGY AND BIOTECHNOLOGY

Although the development of synthetic chemistry and its subsequent impact on our day-to-day lives offers a glimpse of what synthetic biology may become, synthetic biology as a discipline owes as much to engineering as it does to the natural sciences. A synthetic biologist seeks to design and build useful biological systems or devices, in a similar way that an engineer might design and build a bridge, airplane or personal computer. By taking individual components, whose structures and functions are well characterised and understood, and then piecing these components together in a logical manner, the building of useful systems or devices becomes easier, quicker and inherently more reproducible. By adopting this design–build–test approach, many of the complications associated with traditional bio-technology approaches, such as those used extensively in the 1980s and 90s, may be circumvented.[14]

Conceptually, the synthetic biology approach is really no different to how one can build a vast array of different structures or objects using LEGO bricks. Using only a small number of bricks one can design and build a vast array of different structures; a house, a boat, a skyscraper or a space rocket. The form and function of the final object is dictated both by the inherent shape, size and properties of the individual bricks and by their arrangement relative to one another. This is in essence the same basic premise as that of synthetic biology. Biological parts such as DNA sequences, or the proteins for which they encode, can be thought of as 'biobricks', which can, at least in principle, be assembled to satisfy a given design, or to elicit a specific behaviour or function.[15]

Like LEGO bricks, it is both the nature of the biobricks themselves and their arrangement within the finished structure, which confers a specific property or behaviour. Thus, it is conceptually possible to take a small number of biobricks and use them to build a vast array of different biological systems, from biosensors to materials, or even intact functional cells. For example, the DNA encoding a biosynthetic pathway, which in turn produces an anti-cancer drug, could be introduced into a single-celled bacterium, such that when that bacterium is grown in a laboratory it produces vast quantities of the anti-cancer drug quickly and cheaply. Alternatively it may be possible to use biobricks to build a virus that can selectively deliver the same anti-cancer drug to cancer cells within a patient's body. What quickly becomes apparent with this approach is that the outcomes of synthetic biology are limited not simply by the technical capabilities of its practitioners but by their creativity and imagination.

The premise of synthetic biology all sounds very straight-forward, biology as LEGO. In reality, our ability to design, as-semble and engineer biological systems lags some way behind our ability to design, assemble and engineer LEGO-based sys-tems. Joining biobricks together is tough, joining them together in a way that is predictable, testable and more often than not gives us the results we were expecting, is near impossible. Biological components are 'squishy', malleable entities, which often only function in the presence of additional agents like water or chemical cofactors. Frequently they do not operate as hoped or expected outside of their natural contexts or environments. In cases where biobricks can be interfaced they often do not communicate with one another as was hoped or expected, or they behave in a manner that is actually at odds with what was intended. Biology, by its very nature, incorporates elements of randomness, an essential property that enables iterative improvement and optimisation *via* Darwinian evolution.[16] Rationally designed biobrick-based systems frequently struggle to reproduce this property and emergence, a key facet of natural biological systems, is lost.

These issues highlight one great limiting factor in our ability to design and build biology; we do not know enough about how biology works to give confidence that the predictable construction and engineering of artificial biological systems can be achieved. This remains the one great stumbling block for synthetic biology, sure we can build simple 'toy' systems, study them, analyse them and learn much about biological parts and how they interact with one another, but the prospect of being able to build, from scratch, a functional living system or a cell, remains the stuff of science fiction.

So why bother? Surely our lack of understanding of the natural world makes the exercise a futile one? Well no, although the idea of building intact biological entities remains some way off we can strive to make simpler, useful systems that can deliver and help to address real world problems. For example, scientists have engineered microorganisms that can make drugs and fuels, can sense and destroy toxins and contaminants in the environment and can aid in the manufacture of useful materials that form the basis of wound dressings or medical devices. By these baby steps we peel back the layers of biology, exposing its inner workings, and in doing so reveal new routes to harnessing its power.

9.4 ARTEMISININ—SYNTHETIC BIOLOGY COMES OF AGE

If synthetic biology is ever to be considered a truly game-changing technology it is clear that it must deliver on its promise of improving human lives. There is perhaps no greater demonstration of the power of synthetic biology to do this than in its potential to yield new medical treatments. One elegant demonstration of SynBio's ability to do this can be seen in the work of the US biochemist Jay Keasling, whose development of synthetic-biology-based methods for the production of pharmaceuticals signalled the arrival of SynBio on the world stage.

In 2003 Keasling and co-workers reported what was to become the poster child of synthetic biology success stories. Keasling's team had succeeded in transplanting a series of genes from the sweet wormwood plant *Artemisia annua*, into the bacteria *Escherichia coli* (*E. coli*).[17] The transplanted genes encoded a significant portion of a metabolic pathway responsible for the biosynthesis of a chemical compound called artemisinin. The pathway encoded for by the transplanted genes was shown to be functional in the modified *E. coli* cells, which were in turn able to produce a version of artemisinin resembling that extracted from *Artemisia annua* (Figure 9.1).

This work was a *tour de force* of synthetic biology, but why was it so significant for human health? Simply put, artemisinin is one of the must potent anti-malarial drugs known, with an almost 100% success rate in the treatment of the disease. Malaria is a serious tropical parasitic infection that is fatal if not treated. In 2015 there were 214 million cases of malaria reported wordwide and over 400 000 deaths. Artemisinin is one of the most effective malaria treatments, due to its ability to kill all forms of

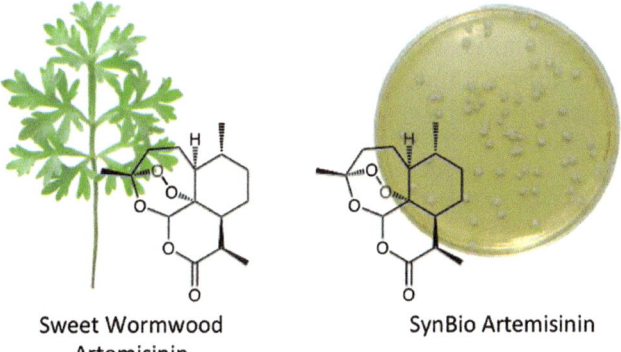

Sweet Wormwood
Artemisinin

SynBio Artemisinin

Figure 9.1 Images and structures of artemisin from sweet wormwood and SynBio artemisin.

Plasmodium falciparum, the causative agent of the disease. It is even active against multi-drug-resistant strains of the parasite. Although artemisinin can be extracted from sweet wormwood plants, the global supply is highly volatile due to fluctuations in the wormwood harvest. For this reason there was a pressing need to develop cheap and reliable methods of producing the drug, to ensure that its global supply could be maintained and that it could be made readily available to all those who needed it.

Keasling's work was a game changer for synthetic biology. It represented the first major synthetic biology breakthrough that had the potential to have significant global impact. The work also highlighted that it was possible to take a significant quantity of DNA from one living organism, a plant, and transfer that DNA to another genetically distinct living organism, a bacteria, in such a way that the proteins encoded by the foreign DNA could be produced *en mass* and in a functional form. This was very much like installing a new piece of software on a computer, allowing the user to perform new tasks or calculations.

Three years later Keasling further extended the capabilities of his technology, successfully transplanting the complete artemisinin biosynthetic pathway into yeast.[18] This work formed the basis of a spin-out company, Amyris, which sought to commercialise the technology, and which has, in turn, started to develop synthetic biology routes for manufacturing a wide range of other useful chemical compounds. Keasling's work captured the imaginations of other synthetic biologists and soon many were attempting to use microorganisms as cell-based 'factories' to produce a diverse range of useful molecules in an inexpensive and highly scalable way. SynBio had not only delivered, but it had delivered big, it had evolved from an academic curiosity into a truly powerful technology.

9.5   HEATH ECONOMICS—REALITY BITES FOR SYNBIO ARTEMISININ

In many respects the story of artemisinin has come to define the burgeoning field of synthetic biology. What was so appealing about Keasling's work was that in addition to its scientific credibility, it hinted at the tantalising possibility of delivering life-saving medicines in a robust and cost effective manner through the rational engineering of biology. Such an approach circumvented the problems of fluctuating sweet wormwood harvests and would act, it was hoped, to stabilise a market that was victim to unpredictable costs and product availability. Translating Keasling's fundamental scientific discovery into a *bona fide* industrial process for artemisinin manufacture proved to be a sizable undertaking. However, in 2014, over a decade after the first of Keasling's SynBio artemisinin papers was published, the pharmaceutical company Sanofi began selling artemisinin that had been, in-part, manufactured using engineered yeast strains. Sanofi used this approach to generate more than 39 million artemisinin treatments, totalling almost 10% of global demand. The early signs for SynBio artemisinin were good but, as is often the case, market forces were soon to intervene. A bumper harvest of sweet wormwood in 2014 and 2015 saw the price of agricultural artemisinin plummet to less that $250 per kg. By comparison, Sanofi's SynBio alternative had a cost price of $350–450 per kg. Simple economics dictated that Sanofi's artemisinin was not cost competitive in a market flooded with cheap naturally sourced material. Other artemisinin manufacturers were also reluctant to buy SynBio artemisinin from Sanofi, due to the competition that this created in an already overcrowded marketplace. In addition, improvements in malaria diagnostics meant a reduction in the frequency of malaria misdiagnosis, and a consequent reduction in the total number of artemisinin treatments required worldwide. Due to the combined effect of these factors, the financial viability of SynBio artemisinin fell. In 2015 Sanofi did not produce any artemisinin using engineered yeast strains, and the company is now moving to sell its manufacturing plant in Italy. Sanofi's goal of producing around one third of the global supply of artemisinin has regrettably to date not been realised.

This all portrays a rather gloomy outlook for SynBio artemisinin. The reality, however, lies somewhere between its hopeful early promise and the economic cold light of day. Sweet wormwood harvests are notoriously fickle, indeed their fluctuation was a major driver in establishing alternative routes to artemisinin manufacture. The recent upturn in agricultural harvests is undoubtedly unsustainable. Inevitably, in the near future, yields of the natural material will fall and as a result the financial viability of semi-synthetic artemisinin will increase. Further-more, new emerging SynBio companies that do not sell artemi-sinin direct to market, but instead manufacture and supply the synthetic precursor to those that do, will be much better placed to gain a foothold in the supply chain, side-stepping many of the complex political and financial obstacles that were faced by Sanofi. Finally, advances in diagnostics and treatment regimes will ultimately reduce the total number of artemisinin treat-ments that need to be produced. From a human health per-spective this represents an excellent outcome, improving prognoses and minimising the risk of drug resistance.

What the story of artemisinin demonstrates is that SynBio does indeed have the potential to deliver new impactful medical treatments, and that it has a major role to play in the future development of medicines and healthcare technologies. The story of artemisinin also highlights the delicate interplay be-tween scientific progress and health economics. Irrespective of the clear medical benefit of a given molecule or therapy, manufacturing that entity at scale, and delivering it to market, is subjected to significant financial and political pressures that extend well beyond laboratory scale discovery science.

9.6   SYNTHIA—SYNTHETIC LIFE IN THE MAKING

If Keasling's work represented a small step forward in the suc-cessful application of synthetic biology, what was to emerge soon after from J. Craig Venter's laboratory was to represent a giant leap. In 2010 Venter and his team published a paper in the journal *Science* reporting the chemical synthesis and implant-ation of an entirely 'synthetic' genome into the nucleus of a bacterial cell.[19] The resulting microorganism's genome was able to decode the information contained within the implanted DNA and quickly adopted the characteristics for which it encoded. If Keasling's artemisinin project was akin to running a new piece of software on a computer, Venter's was analogous to replacing the hard drive. By including distinctive barcoded features within the synthesised genome, sequences of DNA that were unique to the new bacterium, Venter was able to demonstrate that the genome could be readily decoded and reproduced and opened up the possibility of introducing a whole range of novel sequences and subsequent abilities into the microorganism. The breakthrough was heralded by some as an example of man-made 'synthetic life', the panacea that Leduc had proposed almost a century earlier. In reality, Synthia, the name by which the bacteria became known, was quite some distance from synthetic life. However, the implications of this work from both a technical and philosophical stand-point were profound.

To produce Synthia, it took the committed efforts of more than 20 researchers working on the project for over 10 years, at a cost of over $40 million. This was science on a mammoth scale. Venter's team synthesised short regions of DNA that together constituted the genome of the bacteria *Mycoplasma mycoides* (*M. mycoides*) in its entirety. *M. mycoides* was chosen as it has the smallest genome, by number of nucleotide bases, of any known bacteria. Next the researchers transferred these synthesised sequences into yeast cells, exploiting their inherent ability to stitch together sections of DNA into larger chunks. These medium-sized pieces of DNA were then transferred into *E. coli* cells, then back into yeast, and this process repeated for three cycles. At each stage larger and larger chunks of DNA were pieced together until a single stretch of over 1 million bases had been produced, which constituted the entire *M. mycoides* genome. This single piece of synthetic DNA contained all the information required to make a *M. mycoides* cell. In addition, the scientists introduced watermarks into this sequence, such that it could be easily identified and distinguished from 'natural' *M. mycoides*. Next, the synthetic genome was transferred into the nucleus, the control centre of any cell where its genetic information is located, of *Mycoplasm capricolum* (*M. capricolum*), a genetically close relative of *M. mycoides*. The transplanted genome could be read and copied by *M. capricolum*, and the bacteria produced only *M. mycoides* proteins. Furthermore the implanted watermarks could be easily identified within the genome following multiple cycles of cell division. The synthetic genome could be read and written.

So, does Synthia really constitute synthetic life? The transplanted *M. mycoides* genome was indeed synthesised in a laboratory using chemical methods. However, the cellular machineries required to decode and translate the genome into proteins, the functional components of cells, were all derived from the host *M. capricolum* cell. In this author's view at least, Venter's Synthia is still some way away from constituting truly synthetic life, certainly if one adheres to the definition proposed by Leduc. It is unquestionably an important step on the journey to delivering truly artificial living systems, and demonstrates a mastery of biology, but it does not in itself constitute synthetic life in the truest sense. It does however provide a vision of what may be possible. Engineered genomes are now within the grasp of synthetic biologists. They can be fabricated and implanted into other host cells. They can be read, they can be written and they can be translated. The next step in the journey is to exploit this technology for the implantation of new, useful functions, into synthetic genomes. For example, incorporating regions of DNA that encode pathways for useful molecules, including new medicines. Can these functions be dialled in at will? Is there a limitation on what can or cannot be tolerated? These are all fascinating questions that are yet to be explored, but they hint at what may be possible.

9.7   SYNTHETIC BIOLOGY GOES MAINSTREAM—SCIENCE FOR THE PEOPLE BY THE PEOPLE

In addition to the scientific advancements that synthetic biology may deliver, the future also promises a cultural shift in how the design and engineering of biology is undertaken, and by whom. In a similar way to how hobbyist computer builders and programmers drove a digital revolution in the USA in the 1980s, so it may be that bedroom biotechnologists, or biohackers, drive the SynBio revolution.[20] Where earlier incarnations of biotechnology were reliant on huge financial investments, workforces and infrastructures, now motivated hobbyists can access and formulate DNA sequences with relative ease. A raft of design tools and software packages are freely accessible on-line, and there are even companies that will do the experimental work for you at the right price. The Californian based company Transcriptic, for example, offers an automated service allowing users to perform their experiments by proxy, through a digital cloud interface that controls a suite of liquid handling robotics.

Such developments are not restricted to the USA. The Biohackspace in London, for example, provides a community research laboratory where enthusiastic amateurs with backgrounds in a broad mix of professions, such as artists, engineers, biologists and programmers can carry out innovative bioscience projects.[21] The laboratory is funded through members' contributions and supporter donations, and all 'Biohackers' are required to adhere to a DIYbio code of ethics. The Biohackspace provides a forum for SynBio-curious individuals to explore and develop synthetic-biology-based technologies and also to engage more generally with the field. Hackspaces are becoming increasingly commonplace and just as tech powerhouses such as Google or Facebook can trace their origins to dormitory rooms or garages, it is possible that the future of biotechnology and medical technologies will be built in hackspaces or DIY SynBio labs. Synthetic biology is a democratised open science, for the people by the people.

One of the most powerful examples of the accessibility of synthetic biology to the uninitiated is the International Genetically Engineered Machine (iGEM) competition.[22] iGEM is an international synthetic biology competition that was originally targeted at undergraduate university students but has now expanded to accommodate high school students, entrepreneurs, postgraduates and even community laboratories and hackspaces. The competition provides a forum for teams of SynBio enthusiasts to develop and test synthetic-biology-based tools and technologies with the aim of addressing real-world problems. To achieve this teams assemble biobricks, either of their own design, or taken from a registry of previously characterised parts, the Registry for Standard Biological Parts, to build useful synthetic-biology-based systems or devices. Every biobrick developed during the competition is made freely accessible to other iGEM team members and, more generally, to other synthetic biologists, to maximise their value and usefulness.

The first iGEM competition took place in 2004, following the initiation of the iGEM programme as an independent study course at the Massachusetts Institute of Technology in 2003. Over the next decade iGEM grew rapidly, capturing the imagination of expert and non-expert synthetic biologists alike. In 2015 the competition comprised 280 teams from over 30 countries. It is operated by the iGEM foundation, a non-profit organisation that promotes education and competition in the advancement of synthetic biology. The foundation also places a major focus on collaboration and has become a paradigm of community-driven science. The success of iGEM provides a framework for other emerging technologies, and presents synthetic biology as an exemplar of a modern outward-facing approach to scientific research. Many iGEM competitors have been inspired to progress their studies in synthetic biology and some have even developed their competition entries to the point of commercialisation.

9.8   ETHICS, SAFETY AND PLAYING GOD

Synthetic biology is a research discipline that touches on many sensitive areas and as a result has attracted some degree of criticism. In particular, synthetic biology has provoked powerful reactions in those who raise concerns about biosafety and security, and those who question the ethical foundations of the approach. Concerns about biosafety and security focus predominantly on the unregulated exploitation of synthetic biological material and the consequence of its release into the environment.[23] Ethical concerns focus on the 'unnaturalness' of the synthetic biology approach and the notion that synthetic biologists through their actions are 'playing God', perverting nature through manipulation of the molecules of life.[24] Synthetic biologists have been, and continue to be, proactive in recognising and addressing these concerns through transparent dialogue with their critics. It is crucial for all parties that these issues are addressed head-on through reasoned, evidence-based, discussion, and are not hijacked by the misinformed.

With respect to issues of biosafety and security, the risks posed by synthetic biology are no different from those of other disruptive technologies. As alluded to earlier, major areas of concern focus on the malicious or accidental generation of synthetic biological systems that are harmful to other living beings or the environment. Public concerns over such outcomes have been voiced and listened to, and there has been significant consultation and dialogue between governments, synthetic biologists and lay audiences. This has resulted in, for example, the formulation of specific legislation relating to synthetic biology research in both the UK and the USA, amongst others.[25]

Synthetic biologists who work within regulated research environments, *e.g.* universities, research institutes or in industry, are required to adhere to stringent health and safety legislation and must abide by strict codes of conduct. This includes detailed criteria for the containment and storage of genetically modified microorganisms, cell lines or associated materials. Non-compliance results in the imposition of harsh penalties, and the risk of the accidental release of synthetic biological material is effectively nil. For SynBio practitioners operating outside legal frameworks, the biohackers or biopunks, such individuals lack the technical capability and infrastructure required to develop harmful or destructive products, and their activities are inevitably always going to be limited to simple 'toy' systems. Despite this, concerns still exist, and in response synthetic biologists are required to adhere to codes of conduct described within extensive Responsible Research and Innovation frameworks established by governmental bodies. These afford an additional layer of regulation that further minimises the likelihood of the accidental or deliberate release of synthetic biological material.

The second major ethical criticism levelled at synthetic biology focuses on the very essence of the approach itself. Is it appropriate for mankind to manipulate the fabric of life to produce artificial biological entities for the gain of humanity? This question touches on sensitive philosophical issues that have emerged, to a large extent, from misunderstanding and misinformation. The human race has been artificially manipulating biology for its own ends for millennia. The selective breeding of plants and animals, practices that have underpinned the provision of food, fuel and building materials, amongst others, since early man first populated the Earth, represent the exploitation and optimisation of biology for purposeful gain. Supermarket shelves are populated with fruit and vegetable products that have been produced using genetically optimised plants, generated through many years of selective breeding for traits that include flavour, fragrance, appearance or hardiness. This type of genetic manipulation is a freely accepted unquestioned practice, but it is in essence no different from the approach of synthetic biology. Both methods seek to harness and optimise the power of natural systems for greater benefit to man. It is interesting to note that plant and animal breeding practises are perceived by many as 'natural', whereas using the tools of synthetic biology, which are inherently more targeted, predictable and rapid, are viewed with scepticism or fear. DNA is DNA, whether its manipulation occurs in a research laboratory or in a farmer's field, there is in reality no difference in the ultimate outcome.

Finally the notion of synthetic biologists as 'creators' has proven a source of some discomfort in theological circles. What is key here is that 'creation' in the sense of synthetic biology amounts to the assembly of biological components to construct a system, cell or tissue, that exhibits some, or all the defining characteristics of life. Synthetic biology is not an exercise in the creation of living matter from nothing. This is an important distinction as creation in the context of synthetic biology is therefore incomparable to creation in a theological sense. In fact, in the synthetic biology approach one could argue that the role of the practitioner is simply to supply the constitutive parts and requisite conditions that permit the emergence of life.

The ethical debate surrounding synthetic biology continues and its exponents must continue to be open and honest in their approach. If synthetic biology is ever to deliver on its potential it must not be derailed by the misinformed, who liken synthetic biology to some form of Frankenstein science. Synthetic biology, like other disruptive technologies, must remain the subject of stringent regulation and monitoring, and by this means risks and concerns about biosafety, security and the ethical implications of the approach can and will be mitigated.

9.9   DARE TO DREAM, WHERE NEXT FOR SYNTHETIC BIOLOGY

Synthetic biology is a field still in its infancy. The pioneering work that has already been undertaken in the discipline captures the imagination and hints at what may be possible in the future. We have only glimpsed the tip of the iceberg, but what may be possible is beyond tantalising. The potential of synthetic biology to not just deliver, but to deliver big, has captured the imagination of many, not just synthetic biologists but politicians, policy makers, industrialists and lay audiences. Synthetic biology is likely to become one of the most transformative technologies of the 21st century. The advances made in the last decade within this emerging discipline are considerable. Synthetic biology is a field limited at least in part by the imaginations of its practitioners and as such there seems little to hold back the tide of the age of SynBio. As noted by Bill Gates "If you want to change the world in some big way that's where you should start, biological molecules". So, where next for this field? What does the future hold for synthetic biology and when and from where are the next major breakthroughs coming?

The most likely immediate successes look to be in the manufacture of useful molecules. Exploiting the biosynthetic potential of microorganisms through the introduction of foreign DNA has already yielded a number of high-profile success stories, perhaps none more so than SynBio artemisinin. In the future the number and types of molecules that can be produced will increase exponentially. As metabolic engineering becomes more robust and reproducible, many more chemically diverse compounds will be manufactured *via* this route. Microbial strains will be developed that can convert simple feedstocks and waste products into useful molecules. Increasingly more complex chemical compounds will become accessible and cell factories will become the favoured method for industrial-scale manufacture.

Health technologies are also emerging as a major area where SynBio can have significant impact. Methods will be developed to combat bacterial pathogens, potentially by exploiting viruses or predatory 'good' bacteria that seek out and destroy infectious agents that are resistant to even our most powerful antibiotics. Related technologies could be used to prevent food contamination or to decontaminate packaging or medical equipment. Synthetic-biology-based diagnostics will become more robust and reliable, enabling the rapid detection of toxic molecules and infectious microbes. This will allow clinicians to make more informed decisions about the most appropriate treatment strategies for their patients, improving outcomes and minimising distress and discomfort. There will, without question, also be economic benefits from the adoption of SynBio, and research in the area is already primed to become big business. In 2015 alone synthetic biology companies raised more than half a billion dollars in investment to pursue projects ranging from the development of novel therapeutics to cell based diagnostics. It seems inevitable that a SynBio company, with the impact and reach of a Google or Facebook, will emerge soon.

9.10 CONCLUSION

Synthetic biology is a discipline still in its infancy, but the early signs are encouraging. It has the potential to deliver a future where the power of biology can be harnessed and directed to deliver new medicines, materials, foods and fuels. Although clearly rooted in the scientific endeavours of the past, synthetic biology truly is a field all of its own. It should not be judged by the successes or failures of previous biotechnologies, and synthetic biologists must continue to work hard to engage with the public, whenever and wherever possible, to disentangle science fact from science fiction. For many, significant ethical concerns exist and the notion of engineering DNA, the fabric of life, for purposeful application sits uneasily. Such concerns must be addressed and, as a community, synthetic biologists must strive to be open and honest, and not shy away from addressing the difficult questions. The early successes of synthetic biology must be shown to be resilient and reproducible. Can the field continue to deliver game-changing breakthroughs, such as artemisinin, or will the field collapse under a weight of expectation? Who is to say, but one thing that is for certain, there are exciting times ahead and maybe after all, life really is what you make it.

REFERENCES

1. F. Leduc, *La Biologie Synthétique*, 1912.
2. J. D. Watson and F. H. Crick, *Nature*, 1953, **171**, 737.
3. F. H. Crick, L. Barnett, S. Brenner and R. J. Watts-Tobin, *Nature*, 1961, **192**, 1227.
4. W. Szybalski and A. Skalka, *Gene*, 1978, **4**, 181.
5. B. Weiss and C. C. Richardson, *Proc. Natl. Acad. Sci. U. S. A.*, 1967, **57**, 1021.
6. F. Sanger and A. R. Coulson, *J. Mol. Biol.*, 1975, **94**, 441.
7. H. F. Judson, *The Eighth Day of Creation: Makers of the Revolution in Biology*, Cold Spring Harbor Laboratory Press, 1996.
8. A. M. Chakrabarty, General Electric Company, *U. S. Pat.* 4259444 A.
9. M. Fumento, *Bioevolution: How Biotechnology is Changing Our World*, Encounter Books, 2003.

10. R. Carlson, *Nat. Biotechnol.*, 2009, **27**, 1091.
11. International Human Genome Sequencing Consortium, *Nature*, 2004, **431**, 931.
12. M. Morange, *EMBO Rep.*, 2009, **S1**, S50.
13. P. Mali, K. M. Esvelt and G. M. Church, *Nat. Methods*, 2013, **10**, 957.
14. C. M. Agapakis, *ACS Synth. Biol.*, 2014, **3**, 121.
15. K. M. Muller and K. M. Arndt, *Methods Mol. Biol.*, 2012, **813**, 23.
16. C. Darwin, *On the Origin of Species by Means of Natural Selection, or the Preservation of Favoured Races in the Struggle for Life*, John Murray, 1859.
17. V. J. J. Martin, D. J. Pitera, S. T. Withers, J. D. Newman and J. D. Keasling, *Nat. Biotechnol.*, 2003, **21**, 796.
18. D. Ro, E. M. Paradise, M. Quellet, K. J. Fisher, K. L. Newman, J. M. Ndungu, K. A. Ho, R. A. Eachus, T. S. Ham, J. Kirby, M. C. Y. Chang, S. T. Withers, Y. Shibba, R. Sarpong and J. D. Keasling, *Nature*, 2006, **440**, 940.
19. D. G. Gibson, G. A. Benders, C. Andrews-Pfannkoch, E. A. Denisova, H. Baden-Tillson, J. Zaveri, T. B. Stockwell, A. Brownley, D. W. Thomas, M. A. Algire, C. Merryman, L. Young, V. N. Noskov, J. I. Glass, J. C. Venter, C. A. Hutchison III and H. O. Smith, *Science*, 2008, **319**, 1215.
20. M. Wohlsen, *Biopunk: Solving Biotech's Biggest Problems in Kitchens and Garages*, Penguin, 2012.
21. https://biohackspace.org.
22. C. Goodman, *Nat. Chem. Biol.*, 2008, **4**, 13.
23. M. Schmidt, A. Ganguli-Mitra, H. Torgersen, A. Kelle, A. Deplazes and N. Biller-Andorno, *Syst. Synth. Biol.*, 2009, **3**, 3.
24. P. Dabrock, *Syst. Synth. Biol.*, 2009, **3**, 47.
25. UK Synthetic Biology Roadmap Consultation Group, *A synthetic Biology Roadmap for the UK*, 2012; Presidential Commission for the Study of Bioethical Issues, *New Directions: The Ethics of Synthetic Biology and Emerging Technologies*, 2010.

CHAPTER 10

Synthetic Biology–Engineering Tomorrow's Medicines

LIZ FLETCHER* AND SUSAN ROSSER

Centre for Mammalian Synthetic Biology, School of Biological Sciences, University of Edinburgh, Max Born Crescent, Edinburgh EH9 3BF, UK
*Email: liz.fletcher@ed.ac.uk

10.1 INTRODUCTION

History was made early in 2016 when a blind woman in Dallas, Texas became the first person to undergo therapy with an emerging technology called optogenetics.[1] The woman suffers from retinitis pigmentosa, a degenerative condition in which the light-sensing cells in the retina (the rods and cones) die: there is no cure. Optogenetics aims to deploy a clever combination of genetic engineering and light to replace the lost function of these cells and to restore sight.

What has this to do with synthetic biology? Well, if successful, it will be the first clinical evidence that it is possible to 'rewire' a dysfunctional cell to cure a disease. The therapy, developed by RetroSense Therapeutics (Ann Arbor, MI, USA), delivers a package of DNA encoding a light-sensitive pigment to the remaining

Engineering Health: How Biotechnology Changed Medicine
Edited by Lara V. Marks
© The Royal Society of Chemistry 2018
Published by the Royal Society of Chemistry, www.rsc.org

healthy cells of the eye (called retinal ganglion cells). These cells read the DNA and manufacture the pigment, which converts impinging light into an electrical signal that triggers the retinal nerve. The aim is to help restore the missing retinal functions and, hopefully, vision.

This pioneering approach is a very simple form of 'genetic prosthesis' for diseased or damaged cells. It represents the first step on a long and exciting journey in applying synthetic biology and precision genetic engineering to medicine and healthcare.

In this chapter we explore the newly emerging field of mammalian synthetic biology as applied to medicine. Synthetic biology marries our understanding of human genetics and cell biology with the design principles of engineering to recreate or build *de novo* the functions of cells and tissues for the diagnosis, prevention and treatment of disease. The chapter does not provide a comprehensive review of the state-of-the-art in the field but rather a flavour of some of the innovative research and exciting possibilities afforded by this rapidly emerging but challenging area of science. It looks at the role that synthetic biology can play in manufacturing more affordable medicines and how it can facilitate the process of discovering new drugs. We explore how a synthetic biology approach could help realise medicine's drive towards personalised medicine, where therapy can be tailored to meet the very specific needs of each individual rather than the 'one-size-fits-all' approach afforded by more conventional medicine. We examine how synthetic biology could provide a radically new perspective on diagnosing and treating disease and how it might be applied to the design and delivery of cell-based therapies. Finally, we look at how synthetically engineering bacteria that colonise us (our microbiome) may benefit human health.

10.2 WHAT IS SYNTHETIC BIOLOGY?

Synthetic biology is defined as the design and construction of new biological parts, devices and systems and the redesign of natural biological systems for useful purposes. It is a rapidly growing field that has captured the imagination of academic and commercial researchers keen to harness its potential. As Chapter 9 by Race outlines, the discipline has been accelerated by technical advances in gene sequencing and gene synthesis permitting faster and cheaper 'reading and writing' of DNA (deoxyribonucleic acid, the basic building block of life) putting the construction of new living cells and systems within the budget and technical reach of researchers.[2]

When thinking about synthetic biology it helps to think about cells—whether they are from plants, bacteria, animals or humans—as miniature information processing systems. Cells detect a wide array of input signals (*e.g.* light, oxygen, metabolites, toxins) and have internal machinery that can both integrate and execute an appropriate response (*e.g.* generate an electrical impulse, switch on or off genes, release a hormone, grow and divide or even die). Synthetic biology aims to take advantage of this system and, using a toolkit of 'standard' components, build new cellular circuits that can carry out processes faster and more efficiently or even carry out completely novel functions (see Figure 10.1).[3,4]

Today, synthetic biologists have a well-equipped tool box of basic components with which to build. For example, they have access to a wide range of 'on' and 'off' switches that can turn on

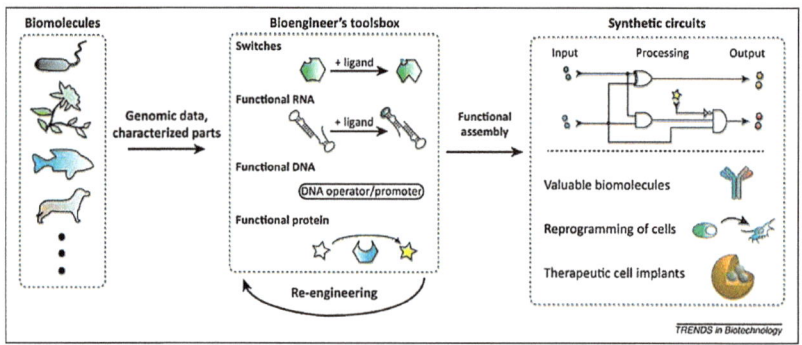

Figure 10.1   The basic principles of synthetic biology. Synthetic biologists look to nature to create a tool box of characterized parts created from DNA, RNA and proteins. These can be reengineered to create additional functions. The parts are then used to create complex synthetic circuits, which provide different types of cellular control for a wide range of medical applications.
Reprinted from *Trends in Biotechnology*, Volume 31, S. Auslander, M. Fussenegger, From gene switches to mammalian designer cells: present and future prospects, 155–168, Copyright (2013), with permission from Elsevier.

or off the genes that produce proteins. These simple components can be further engineered to respond to different 'inputs' (*e.g.* a drug) or to generate more subtle 'outputs': for example an output might be conditional on the presence of two or more inputs, or could be amplified in the presence of a second signal. It is now possible to build very sophisticated circuits that can, for example, create oscillations, tuneable outputs or even counting devices in cells. The time and effort involved in the 'trial and error' of iteratively building and testing new synthetic cellular circuits is likely to be reduced in the future through the rise of automation and robotics in research laboratories. Only imagination, manpower, time and budget limit the possibilities.

10.3 MOVING ON FROM MICROBES

The early 'wins' for synthetic biology have come from re-engineering industrial microbes such as yeast (*Saccharomyces cerevisiae*) and bacteria [*e.g. Escherichia coli* (*E. coli*)] to make useful chemical building blocks for bioplastics, biofuels, medicines (*e.g.* the painkiller hydrocodone) and even textiles (*e.g.* synthetic spider's silk is now being made in yeast by Bolt Threads, a biotechnology company based in Emeryville, CA, USA). These industrial microbes are living 'cellular factories' harnessed to make materials faster, cheaper and potentially more sustainably than traditional methods.

Many of the early applications of synthetic biology were relatively simple modifications of microbial cellular pathways; for example, to encourage a cell to increase the yield of a natural metabolite or to enable it to live on a different feedstock. Today, entire metabolic pathways have been reconstructed *de novo* using a tool box of basic synthetic biology building blocks. One of the most highly publicised successes is from the lab of Jay Keasling and co-workers (University of California, Berkley, USA) who reproduced the synthetic pathways for the anti-malarial drug artemisinin, traditionally derived from the Chinese wormwood plant, in yeast. This drug is also explored by Race in Chapter 9.[5,6] More recently Christina Smolke and colleagues at Stanford University have managed to transfer the genes encoding the pathways for opioids from plants into yeast, again potentially helping to reduce the need for plants as a source of this useful pain killer. Synthetic biology provides new routes for producing commercially valuable speciality chemicals and medicines.

Applying synthetic biology to mammalian cells poses some significant additional hurdles: human cells are far more complex than yeast and bacteria, which complicates the ability to define standard genetic components and the rational design of new pathways; the mammalian synthetic biology 'tool box' is currently rather bare, which limits what can be constructed; there is also no 'standard' mammalian cell on which all researchers can work, which poses headaches when interpreting the consequences of implementing what is learnt about engineering one type of cell (*e.g.* kidney) on another (*e.g.* liver); and finally, most human cells are, of course, components of complex multicellular structures such as tissues and organs, which involves highly sophisticated levels of interaction across space and time.

All in all, mammalian synthetic bioengineers have quite a task ahead, but they are lured by the appeal of using the exquisite sensitivity and responsiveness of the mammalian cell to design and develop new strategies for diagnosing, preventing and treating disease. One of the most commercially attractive applications, and one perhaps closest to being realised, is the application of synthetic biology for the production of medicines and vaccines.[7,8]

10.4 SYNTHETIC BIOLOGY FOR MANUFACTURING MEDICINES

The cost of medicine globally is around a trillion dollars and rising, driven by a growing global and 'greying' population, the rapid spread of drug resistance and emerging new infections (*e.g.* pandemics like Ebola and the Zika virus), and the high cost of managing complex, chronic diseases such as obesity and diabetes. Synthetic biology could offer new tools to expedite and reduce the cost of the manufacture of medicines and help to make them accessible for many more patients.[9] It also could enable us to move more quickly towards realising personalised medicine, which recognises that there are individual variations in both underlying disease and in response to therapy and aims to tailor treatment to fit.

10.4.1 Better Biologics

Biologics are typically drugs that are made from large protein-based molecules. They include treatments that use hormones (*e.g.* insulin, growth hormone) and artificial antibodies known as monoclonal antibodies (*e.g.* the anti-arthritis drug Remicade). Such drugs mimic or replace natural proteins and, as such, are potent and highly effective medicines that have been a great success story for 21st century healthcare, currently making up a staggering 40% of all drugs sold today.

The downside is that these structurally complex proteins cannot be made by conventional chemistry. As Chapter 2 outlines they must be manufactured in a living cell such as Chinese Hamster Ovary (CHO) cells (these are an immortal tumour cell line) or *E. coli* (a common bacteria found in the gut). Living cells need rather more nurturing to be productive 'factories' for biologics, the process takes time, yield can be low and it can be tricky to isolate the final protein: as a result, the end product can be expensive. In some instances, the yield of a protein is simply too low to make the product commercially viable. Overall, the real cost is a human one as many individuals and healthcare systems simply cannot afford these medicines.

With increased understanding of the process by which a cell manufactures these proteins, synthetic biologists globally are devising new ways to encourage these cells to work harder and more predictably. They can now engineer CHO cell lines to have improved metabolic performance so that they can make more protein under the same conditions, which should improve yield. Others are engineering cells to change the way the proteins are processed (such as adding or deleting sugars or lipids) to help modify the biologic's stability or its duration of action in the body, thereby making it more clinically useful.

Synthetic biology is also being used to engineer cells to make novel biologics that are not found in nature but could make superior medicines: for example, bi-specific antibodies, which are capable of binding simultaneously to two different targets and redirecting immune cells to destroy tumours, or single-chain antibodies that are smaller than normal antibodies so better able to penetrate tissues to access diseased cells. All these manipulations could, ultimately, help improve access to biologics previously unavailable to patients on the basis of cost.

Others are of the view that we should simply remove the need for the cell at all. Several labs, including that of Michael Jewett (Northwestern University, Evanston, IL, USA), are developing cell-free systems, in which bacterial extracts are purified to extract the vital parts of a cell's protein production machinery and the cell itself removed. Having the production systems cell-free potentially makes for a more robust, more tuneable and more potentially scalable system to produce engineered proteins in bulk.[10] Although this idea is some way from being applied commercially, in theory at least, this could be a robust method to produce a wide range of biologics and potentially even deliver affordable 'personalised' biologics tailored to meet the specific disease profile of an individual.

10.4.2   Synthetic Vaccines

Another important class of biologics are vaccines, which play a vital role in protecting both humans and livestock against infectious disease. Most vaccines, until very recently, have been made by identifying and purifying the pathogen in question, rendering it non-infectious, growing this in cultured cells or chicken's eggs and then harvesting it. As Buckland shows in Chapter 3, biotechnology has already begun to change this process. Yet, making a vaccine can still take months—far too slow to protect against rapidly emerging and fast-spreading infections such as influenza and Ebola.

Today the advent of affordable and fast genome sequencing and DNA assembly allows for high-throughput construction of prototype DNA vaccines within days rather than months and could even circumvent the need for the infective virus at all. In 2013, US biotech company Synthetic Genomics (La Jolla, CA, USA) in collaboration with Novartis Vaccines and Diagnostics and the J Craig Venter Institute reported that they had managed to construct a prototype synthetic flu vaccine within a few days, rather than weeks. They downloaded the gene sequence of the flu virus from the internet and constructed and validated artificial genes suited for stimulating an immune response. The selected genes were then inserted into an existing non-infectious viral 'skeleton' ('vector') and the resulting synthetic virus manufactured in cultured cells.[11,12] Later the synthetic viral vaccine was converted into a vaccine suitable for clinical trials in volunteers, with promising results. In principle, this strategy could be applied to develop a vaccine against any emerging virus for which the gene sequence is available with the speed necessary to stop it spreading.

Synthetic biology could eliminate the need for the viral vector completely. Enrico Mastrobattista and colleagues at the University of Utrecht (The Netherlands) are using the DNA coding for pathogen antigens (*i.e.* proteins seen by the host as 'foreign') and an adjuvant (a substance that is added to a vaccine to boost the immune system's response to the vaccine) and packaging this with the bacterial machinery needed to 'read' this viral DNA within a tiny sac of fat molecules called a liposome. The DNA-carrying liposome can be injected and it works rather like an artificial virus to stimulate the host immune system.[13] This simple and relatively cheap and reproducible platform could be used for swift vaccine production to respond to a fast moving epidemic and at a scale that could also make viable the demand for vaccines for individual patients.

10.4.3 Engineering Cells for Drug Discovery

At the last estimate it takes more than \$2 billion and 10–15 years to bring a new drug to the market: an eye-watering investment that is elevated not just by the rising cost of research and development and regulatory approval but the rather dismal success rate of actually spotting and developing effective new medicines. Moreover, unanticipated toxic effects of drugs on the liver and heart mean that many approved drugs have to be withdrawn even after they have been approved and gone on sale. Another as yet imperfectly understood issue is the great variation in individual responsiveness to the same drug: some patients respond well while others may not respond at all. The great technical advances of the past decade—whole-genome sequencing, automation and robotics for high-throughput screening of drugs and *in silico* screening—have done little to solve these problems. Synthetic biology could make feasible novel and more cost-effective ways of carrying out drug discovery[14] and move us further towards making personalised medicines a reality for tomorrow.

10.4.4 Disease in a Dish

Often the first time a potential new drug candidate is tested on a human is in phase 1 (early) clinical trials on healthy volunteers. Before that a variety of animal, or animal-derived, cell-based tests are used to assess if a drug is effective and safe. While animals with the relevant disease are often used later in the drug development process such tests are expensive, time consuming, ethically questionable and often poorly representative of the human pathology. Indeed, for some product types and in some regions (*e.g.* cosmetics in the European Union) animal testing is banned. A radically different approach has been proposed that marries synthetic biology, precision genetic engineering and advances in stem cell technology to create more accurate and representative cellular models of human disease that can be used to identify promising new drugs.

Stem cells provide an excellent source of drug-testing material, being the source of potentially endless supplies of cells of all types. With the right chemical stimuli in place, human embryonic stem cells can be coaxed into proliferating into any cell type (so-called 'primary' cells) that could provide more human drug screens. For example, an embryonic cell can be chemically encouraged to form heart cells under one set of conditions or liver cells under another. They can also be encouraged to generate millions of progeny, thereby providing sufficient material for industrial-scale high-throughput screening.

The use of embryos clearly poses ethical and regulatory issues and the supply of cells from this source are limited. One solution was developed by Shinya Yamanaka and a team at Kyoto University. Around a decade ago they developed a method of reprogramming adult cells (usually skin-derived fibroblasts) into a more youthful state by encouraging them to switch on several genes key to development. Bathing adult cells in a chemical cocktail encouraged the adult cells to revert back into a foetal-like state and then, much like embryonic cells, they could be induced chemically to differentiate into any desired cell type (see Figure 10.2). Such cells are known as induced pluripotent stem cells (iPSCs).

The unique quality of iPSCs is that they retain the genetic profile of the individual from which they are derived: hence

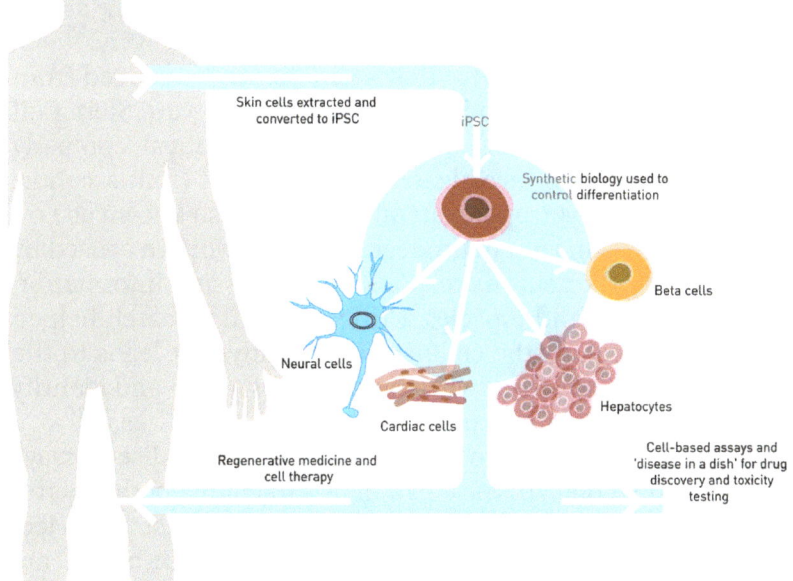

Skin cells extracted and
converted to iPSC

iPSC

Synthetic biology used to
control differentiation

Beta cells

Neural cells

Hepatocytes

Cardiac cells

Regenerative medicine and
cell therapy

Cell-based assays and
'disease in a dish' for drug
discovery and toxicity
testing

Figure 10.2 The production and application of induced pluripotent stem cells (iPSCs) in medicine. Adult skin cells can be directed towards a state from which they can be converted into a wide variety of cell types for use as drug discovery tools or for human therapy. Synthetic biology can be used to provide better control over the types of cells generated and to provide additional ways of measuring the efficacy and toxicity of drugs in cell-based assays.

iPSC-derived cells created from a patient with a particular genetic disorder continue to carry that error. Therefore iPSC technology opens the door to highly relevant assays for human disease, useful for drug screening and gaining new insights into the mechanisms of a disease. By using precise genetic engineering and synthetic biology, researchers can start to interrogate disease pathways and engineer a range of useful 'readouts' that could facilitate better understanding of the nuts and bolts of the disease processes at the molecular level. For drug developers such work could provide a deeper and more nuanced insight into why a drug could, for example, potentially damage the liver or heart, both common reasons why otherwise very beneficial drugs are withdrawn from the market. This concept is not so far from reality: major pharmaceutical companies are already adopting iPSC for both understanding disease and for drug development.

What might seem like science fiction is in fact making a profound change in the way we think and do drug discovery.[15]

Some idea of how powerful iPSCs can be can be gauged from the work of Kevin Eggan and Lee Rubin of the Harvard Stem Cell Institute (Cambridge, MA) with the pharmaceutical company GlaxoSmithKline who investigated iPSCs derived from a cohort of patients with amyotrophic lateral sclerosis. Also known as Lou Gehrig's disease, this is a neurodegenerative condition caused by the death of motor neurons responsible for passing information between the brain and muscles. Critically, their research helped establish more clearly the cellular abnormality that leads to the disease and provided them with the means to screen and identify new agents to control this potentially devastating disease.[16]

Ultimately the hope is to create bespoke personalised assays for an individual (aka 'disease in a dish'), to enable both a better understanding of the patient's own disease and to test and select the most suitable treatment. This could help reduce healthcare costs by eliminating the prescription of ineffective medicines and helping to limit adverse side effects that are so frequently experienced.

10.4.5   Novel Antimicrobials

Where synthetic biology could prove particularly useful is in the development of new antimicrobials.[17] Drug development cannot keep pace with the natural ability of bacteria to evade antimicrobials and the spread of multi-drug resistance poses a serious challenge to modern medical practice. Antibiotic resistance is now considered a global threat and so warrants special attention.

Part of the problem for drug discovery is finding efficient ways to screen for new antibiotics. A vital characteristic of a safe but effective antimicrobial agent is that it must kill harmful bacteria but preserve host (human) cells. Martin Fussenegger and colleagues at the University of Basel and ETH Zurich developed a clever way to screen for drugs with this power of discrimination: they engineered mammalian cells to carry a bacterial gene encoding a known target for an antibacterial compound. This was done by inserting a genetic circuit that replicated a key bacterial growth controller linked to a detectable probe, such as a chemical or enzyme; cells can be screened with small molecules to identify those that are both non-toxic to the cell and can cross the membrane of the mammalian 'host' cell and bind to the bacterial target. This is a robust platform that can be applied to screen a wide range of potential bacterial targets for novel anti-microbial agents that might clear infection or, alternatively, overcome existing antibiotic resistance mechanisms.[18] In principle, you could systematically engineer, test and screen for drugs to kill a range of pathogens.

Synthetic approaches can also be applied when searching for novel chemicals with antimicrobial properties. With the rapid advances in DNA sequencing it has been possible to read the genetic code of thousands of bacteria and fungi, which are known to be rich sources of antimicrobial agents (*e.g.* penicillin was originally sourced from a fungus). However, many potentially useful microbial extracts are simply not detectable because they are either present at minute levels or the genetic pathways that encode them are silenced in the cell. Researchers can overcome this problem by using synthetic biology to identify and recreate *de novo* the biosynthetic machinery of even 'silenced pathways' and to insert them into microbes for manufacturing the natural extract. This could be done for millions of genetic pathways in a high-throughput manner, offering a very rich source of novel chemicals with potential antimicrobial activity. Some success has been achieved with this approach by Bradley Moore and colleagues at the Scripps Institute in San Diego (CA, USA), who successfully 'reactivated' a cluster of genes that make a novel antibiotic called taromycin A.[19]

10.4.6 Mini Organs for Better Toxicity Screening

Finally a word on structure. Even the best cell-based drug assay has its limitations as the ultimate targets for any medicine are the tissues and organs of the body, which are complex 3-D structures which are influenced by multiple and complex chemical and mechanical forces. Research is rapidly moving towards imitating the exquisite three-dimensional architecture of human tissues and organs and using them for *in vitro* screens for drugs in development.

Tissue engineers are using synthetic biology to both understand the complex morphology of tissues and organs and to build a toolkit of cellular 'controllers'—simple cues that instruct, for example, a cell to grow, divide, form layers or connect with other cells. These can then be built into cells to create novel mini-organ structures called 'organoids', which have a similar composition and structure to the natural tissue or organ just on a smaller scale. Several non-engineered 'mini' organs have been grown so far—liver, thymus, kidney and even parts of the brain. These structures better replicate the natural state of cells in the tissues of the body—with the juxtaposition of many different cell types in highly organised three-dimensional structures; as a result they make for more representative drug screens than the homogenous cell cultures more traditionally used in high-throughput drug screening paradigms. Hopefully these mini-organs can weed out potentially problematic drugs before they reach patients.

Ron Weiss (MIT, Boston) and Jamie Davies (University of Edinburgh, UK) among others are working on creating synthetically engineered tissues, building in additional levels of subtle control on structure (so-called synthetic morphology). Synthetic biology could be used to instruct cells to assemble as a result of a chemical (*e.g.* a drug), to form defined patterns and shapes, to change structure over time and even to 'age' on cue.

Some of this research has started to be developed commercially. One US-based company, Emulate (Boston, MA), founded on pioneering research in the field by Don Ingber of the Wyss Institute (Harvard, MA), has developed a range of 'organs-on-a-chip' products, where different cell types are reconstituted into sophisticated structures that can be readily monitored by their supporting 'chip.' They have miniature models of beating hearts, breathing lung tissue and even bacteria-coated gut walls. In the future, we may be able to eliminate the need to carry out any drug testing on animals, replacing them with superior human-like assays, tailored using synthetic biology, which benefits us all.

10.5 SYNTHETIC CELLS AS DIAGNOSTICS AND THERANOSTICS

Healthcare today is slowly moving towards maintaining 'wellness' and prevention rather than the more traditional diagnosing and treating late-stage disease. The theory is that if we could foresee the warning signs of disease and intervene early with a targeted and individually tailored treatment we could radically improve the lives of people and reduce healthcare costs. Based on this a new concept has been developed—theranostics. This is a way of combining a diagnostic and a therapy in the same device or agent. Although most of the examples of cellular theranostic devices are still very experimental, they provide compelling evidence of the possibility of putting theranostics into medical practice.

10.5.1 Cellular Prosthesis

Many human diseases cancers, allergies, diabetes, obesity, epilepsy, Parkinson's disease—arise from cell dysfunction. In principle at least, the missing (or aberrant) cellular process could be corrected using a synthetic (prosthetic) circuit. Ideally the circuit should be controllable by an external signal (*e.g.* a drug, light, heat) so that the prosthetic circuit can be switched on or off to produce the necessary therapeutic output (see Figure 10.3).

One of the leaders in the field, Martin Fussenegger, first designed a prosthetic circuit for treating diabetes using light as a

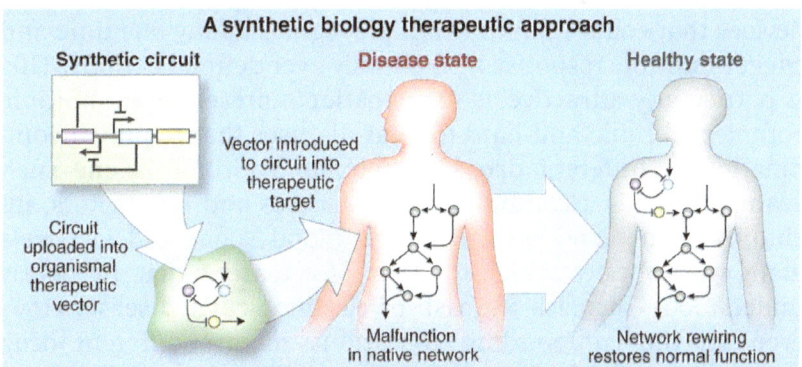

Figure 10.3 The concept behind synthetic prosthetic circuits is to fix cellular malfunctions associated with human disease. The prosthetic gene circuit is inserted into a cell and loaded into the patient to replace the missing function.
From W. C. Ruder, T. Lu, J. J. Collins, *Science*, 2011, **333**, 1249. Reprinted with permission from AAAS.

trigger. This he did in 2011 using mice. He used a native retinal photopigment (called melanopsin) and genetically rewired its internal circuitry—using synthetic biology—so that instead of generating an electrical signal when illuminated it turned on the expression of a novel gene. In this instance the gene was for glucagon-like peptide (GLP), which controls blood sugar concentrations. A flash of blue light activated these engineered cells and stimulated them to manufacture and release GLP. When these cells were encapsulated in a biocompatible and semi-permeable material, and implanted just under the skin of mice suffering from Type 2 diabetes, blue light triggered the release of GLP and helped reduce dangerous rises in blood glucose.[20] In this way a simple cell was successfully reprogrammed to cure a metabolic disease.

Since then, researchers have built a wide array of synthetic circuits and engineered them to be regulated by all manner of stimuli, ranging from light and chemicals to electromagnetic forces. In the latter case the researchers engineered a human cell to carry tiny iron particles that warm up when exposed to radio waves and trigger a cascade of cellular processes that ultimately lead to the release of insulin.

Synthetic biology provides an opportunity to build in lots of layers of sophisticated cellular control just as is possible with an electrical circuit or microchip. In theory you could build cellular devices that could monitor multiple signals at any one time and trigger multiple responses, potentially over defined periods. This is particularly attractive as many patients present with multiple complex, chronic and hard-to-treat diseases that require a combination of different drugs. Metabolic syndrome is one such example: this is a combination of diseases and risk factors, including high blood pressure, high blood sugar and fat levels along with obesity, which together—for reasons that are poorly understood—increase the risk of cardiovascular disease. However, all the confounding conditions require different drug treatments and this, in turn, creates complications (and cost) in treating patients.

A clever solution implemented by Fussenegger and his team in 2013 was to create a synthetically engineered circuit that triggers the production *in vivo* of therapeutic agents for two disorders in response to administration of a single approved medication.

A prototype system was built around the finding that a drug approved to treat high blood pressure (guanabenz) could bind to a cellular receptor not linked to its original therapeutic action. This receptor, and the cellular pathway it activates, was then re-engineered to activate, in turn, two new synthetic gene circuits, one that resulted in the expression of GLP protein (to control blood sugar) and the other a protein called leptin (to control body fat). This 'three-in-one' system worked successfully in mice developing symptoms of metabolic syndrome to bring down blood pressure, blood fat, blood sugar and body weight.[21]

10.5.2 Theranostics on the Drawing Board

The circuits described previously offer tantalising evidence of the possibility of creating a finely tuned, living, drug delivery system. However, ideally we need such a system to be able to independently monitor for the early signals of disease or dysfunction and respond promptly with the appropriate therapeutic response, adjusting as necessary. In other words we need to engineer a 'closed loop' prosthetic circuit.

One of the first examples of the 'closed circuit' was made by Fussenegger in 2010. He engineered an elegant cellular device that could maintain homeostasis of uric acid in the blood stream. Uric acid is a natural chemical produced by the breakdown of purines that are found in food and drink such as liver, beer, beans, peas and anchovies. If levels rise too high (*e.g.* through diet, genetics, drug treatment or following cancer therapy) urate crystals form in the joints and kidneys result in a painful condition called gout.

The system comprised a human urate transporter (that monitors and takes up uric acid from the blood stream) and a genetic 'thermostat' built into a human cell line (called *HeLa* cells, an immortal human cell line). The circuit was constructed so that rising levels of uric acid inside the cell switched on expression of a gene that makes urate oxidase, an enzyme that converts uric acid to harmless allantoin. The levels of uric acid fall, the thermostat is turned down, less enzyme is produced and then uric acid levels start to rise again triggering the cycle to repeat. Ultimately a steady state is achieved in a culture of synthetically engineered cells. To see whether this could work in a living system, Fussenegger placed the cells into protective capsule and implanted this into mice. The cells still did their job—as uric acid levels in the mouse rose, the cells kicked in to restore the balance.[22]

The repertoire of synthetic cellular 'theranostic' devices has been expanding quickly. Although very much still in the research phase, there is early but encouraging evidence of the potential for these cellular devices for treating other 'hard to control' diseases including hypertension, obesity and diabetes. Experimental cellular systems have been shown to protect against liver damage, to control insulin release in Type 1 diabetes, to control flare ups of the skin disease psoriasis and even to fine tune the timing of artificial insemination of cattle. The possibilities are limited only by the building blocks available to cellular engineers, imagination and clinical need.

10.6 ENGINEERING CELL THERAPY

A cure for many diseases can only be achieved by repairing or regenerating diseased, damaged or dysfunctional tissues or organs. Traditionally this has been done through cell or tissue transplantation from human donors, but the demand is now no longer met by supply.

Developments in stem cell technology (as described above) have opened the door to a new source of potentially endless supplies of transplantable material. Such cells can proliferate and differentiate into whatever transplantation cell type is needed, they can migrate to diseased or damaged areas and they can integrate with healthy tissue. However, stem cells do come with 'behavioural problems'—they can migrate to places where they are not needed, start to proliferate in an uncontrolled manner (forming a tumour) or simply just die. This has been something of a barrier to the more widespread application of cell therapies and regenerative medicine.

10.6.1 Building in Controls

Results of early studies have indicated that combining synthetic biology and stem cell technology might be a route to asserting the necessary checks and balances on stem cells to ensure they will be safe and predictable cell therapies. For example, synthetic biology could be used to insert a genetic 'killer switch' in a stem cell to trigger its death (a process called apoptosis) after it has divided a sufficient number of times to repair a tissue or replace a lost function; this would reduce the risk of tumour formation. These killer switches could be engineered so that they could be turned on by external signals such as an approved medicine or agent like biotin (part of the vitamin B complex). In that way, clinicians could still exert control over the cells long after their transplantation.

Another opportunity for engineering control into stem-cell technology is being able to control what type of cell they turn into more predictably. Researchers such as Steve Pollard (University of Edinburgh, UK) are working to insert circuits containing native, or synthetic, so-called 'master control genes' that mimic what happens in nature to drive the development of the embryo into a mature adult. This is not trivial, as the process of stem cell differentiation is far from perfectly understood, but, through trial and error, progress is being made. In one recent study, a designer network controlling the switching on or off of three key transcription factors could programme human iPSCs into glucose-sensitive insulin-secreting beta cells of potential value for transplantation.[23] Derrick Rossi and coworkers (Harvard Medical School) showed that synthetic RNA can also be used to control the type of cells a stem cell differentiates into, circumventing the potential problems inherent in inserting 'foreign' DNA into stem cells destined ultimately for therapy.[24] Some of his research is now being developed by the biotechnology company Moderna Therapeutics (Cambridge, MA).

In the future a combination of stem-cell technology and synthetic biology approaches could ensure a predictable, reproducible and safe source of cells for therapy and regenerative medicine.

10.6.2 'CAR-T Blanche' for Cancer Therapy

Where synthetic biology and cell therapy have advanced most in application is in the generation of personalised immune-based therapies for cancers. Known as immunotherapy, such treatment harnesses the patient's own immune cells to attack and clear their cancer. As Chapter 5 shows immunotherapy has undergone slow but steady progress over the last two decades. Where synthetic biology is currently generating particular excitement is in designing and building 'turbo charged' immune cells for chimeric antigen receptor (CAR) therapy (CAR-T). This treatment was pioneered from the late 1980s by Zelig Eshar (Weizmann Institute of Science, Israel), Steven Rosenberg (National Institute of Health, USA), Stanley Riddell (Fred Hutchinson Cancer Research Centre, Seattle), Carl June (University of Pennsylvania) and Michel Sadelain (Memorial Sloan Kettering).[25,26]

The particular components of the human immune system that the CAR-T system works with are T cells. Such cells are capable of spotting and killing aberrant cells, but cancer cells learn to evade or suppress these immune cells, leaving the tumour to grow unchecked. CAR-T aims to address this issue. In CAR-T, the patient's T cells are harvested from their blood and then genetically modified using a safe virus to incorporate a synthetic circuit encoding synthetic CAR receptors that can detect and bind a protein (the antigen) expressed by the patient's specific cancer (see Figure 10.4). (For example, in the blood cancer called acute lymphoblastic leukaemia an antigen called CD-19 is commonly expressed.) The newly engineered T-cells manufacture and express the cancer-seeking receptor on their surface so that when they are transfused back into the patient they 'spot' and eliminate the cancer cell.

Several biotechnology companies—Juno Therapeutics, Kite Pharma, Cellectis and Bellicum Pharmaceutical—are developing this technology for patients. They have attracted substantial private investment, reflecting, in part, the unprecedented results seen in early clinical trials where more than 90% of patients responded to treatment and a majority were cured completely.[27] The benefits also appear to be durable as the 'super' T-cells can proliferate following transplantation and so can provide protection against the cancer recurring.

Further genetic redesign of these T-cells may make them both more potent and more selective: T-cells can be prone to so-called 'on-target, off-tissue' effects, where they non-discriminately hit all cells bearing the 'target' antigen and not just the cancer cells. Several groups have come up with a clever solution in the design

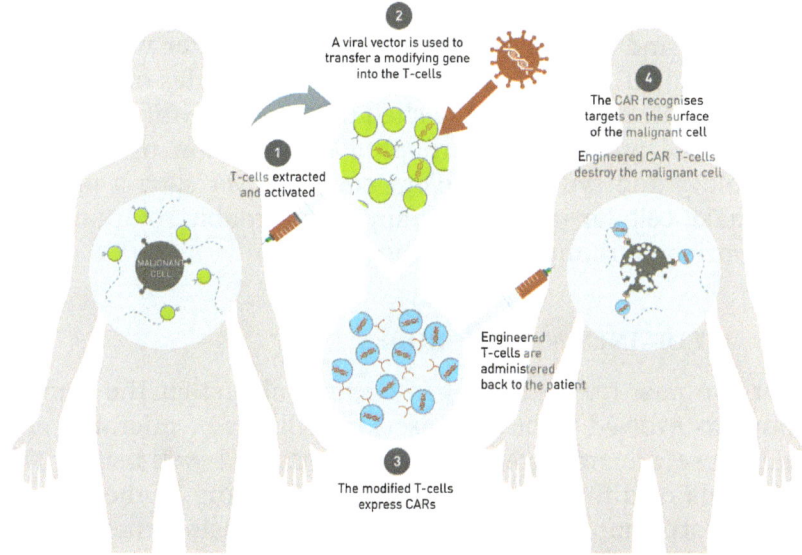

Figure 10.4   The basics of CAR-T therapy. The patient's T cells cannot lock onto the cancer cells, which continue to proliferate. Blood is taken and the T-cells extracted. These are then infected with a gene construct that the cell then converts to the required CAR receptors. The T-cells are grown and re-infused into the patient where they can now lock on and destroy the tumour cells.

of T-cells expressing CARs for two antigens found on the cancer cells: T-cells are only activated when both cancer antigens are present but not (*e.g.* on a healthy cell) on cells bearing just one. An alternative is a system where one CAR is triggered by the antigen and the other CAR by a drug so that the clinician can control the timing of the activation of the T-cells. Others are exploring ways to arm T-cells with means to encourage their own proliferation to overcome cancer's ability to repress the activity of these cells. There is also important ongoing engineering work being done to 'tune down' their activity to help reduce the risk of what can be a dangerous side effect of CAR-T called a cytokine storm, where the destruction of the cancer cells triggers a powerful immune response in the patient, which can be fatal.

The beauty of CAR-T is that it could, at least in principle, be adapted to address any individual's unique cancer, offering a truly personalised and highly specific therapy not possible with current 'gold standard' cancer treatments such as chemotherapy and surgery. The caveat is that CAR-T is an intensive bespoke procedure and will be prohibitively costly to offer widely. One solution might be to generate unlimited quantities of synthetic 'universal' T cells as an off-the-shelf therapy compatible with any patient and any cancer. Perhaps a combination of iPSC cell technology and synthetic biology could deliver such a product and make cellular immunotherapy an affordable solution for a wide variety of immune-based disorders.

10.7 MEDICINES AND MICROBES

Over the past few years it has become clear that the human body has evolved to be colonised by trillions of microbes and that these are important to our health and well-being. The hundreds of different types of bacteria lurking in the human gut perform important roles, including releasing energy from food, metabolising bile acids, fats and drugs and synthesizing life-maintaining molecules like vitamin B and vitamin K. While it is self-evident that 'bad' bacteria cause gut problems (such as diarrhoea, irritable bowel and Crohn's disease), it is less intuitive that gut bacteria might contribute to the development of diabetes, obesity, allergies, cancers and even, as proposed by some, neurological disorders such as autism and depression. While such ideas require further substantiation, there is much excitement around the gut microbiota (the collection of microbes) as a radical new target for therapeutic intervention.

Several synthetic biologists are engineering bacteria to sit in the gut and perform the role of living drug delivery systems. Commensal gut bacteria have been engineered to secrete therapeutic agents such as insulin for diabetes, peptides to treat HIV and immune-modulating agents. Jim Collins and Timothy Lu (both now at MIT) took a different route and engineered *Escherichia coli Nissle* (a 'good' bacteria that is commonly found in the gut and in probiotic food products) to help break down excess levels of metabolites that are produced in some rare genetic disorders such as urea cycle disorder and phenylketonuria. They inserted genetic circuits into *E. coli* that enable them to sense a patient's internal state and switch on or off a metabolic pathway to remove (or increase) levels of the metabolite as the bug moves through the gut: in effect, this could deliver the right amount of therapy at the right place and at the right time. Their technology is currently being developed by a biotech start-up they helped found called Synlogic (Cambridge, MA).[28]

Alternatively, it is possible to employ bacteria as guardians against disease. Bacteria communicate with one another through a chemical language called quorum sensing, a system that bacteria use to co-ordinate certain behaviour. These chemical signals accumulate as the bacterial population grows and instruct bacterial colonies to change their behaviour. This includes forming an antibiotic-resistant biofilm (a particular problem with orthopaedic implants) or increasing the virulence of the bacteria. With synthetic biology it is possible to introduce 'friendly' bacteria synthetically modified to manufacture an agent that can neutralize this bacterial cross-talk and mitigate the damaging consequences of bacteria in the body.

Others are taking a different approach, David Bikard and Luciano Marrafini (Rockefeller University) and Xavier Duportet and Timothy Lu (MIT) engineered synthetic bacteriophages (phages) to selectively target and wipe out harmful bacteria in the gut microbiome. Phages are bacterial viruses and infect, replicate and kill bacteria. The new phages are engineered to carry synthetic constructs and genetic editing machinery that is designed to selectively wipe out the target bacteria—creating a highly selective antibiotic.[29] Start-up company Eligo Biosciences (Paris, France) is developing the technology and has prototyped a way to produce the phage quickly and in large numbers so they can develop highly specific antibiotics against a range of pathogens.

Only time will tell whether 'microbiome engineering' is feasible *in situ*, as the human microbiome remains a complex and poorly understood system. It is not clear whether it is possible to 'repopulate' the gut with synthetically engineered microbes, given that they will face stiff competition from native bacteria that have evolved to survive in the human gut. However, as we learn more about the microbiome on many surfaces of the human body it seems likely that they will attract the attention of the synthetic bioengineers keen to restore bacterial harmony and health.

10.8   CONCLUSION

Synthetic biology offers no immediate panacea for human disease and disability, but the foundations are now laid for new strategies for diagnosis and treatment not even dreamed of just a decade ago. The most immediate benefit of synthetic biology will be in its ability to develop the tools, technologies and new ways of thinking researchers need to gain deeper insight into the basic of human cell biology—effectively 'learning by building.' In turn, this will expedite fundamental biological and medical research and drug discovery with broad and far-reaching benefits for medicine and healthcare globally. We can be confident that today's bioengineers will find ever more creative means of controlling and applying synthetically modified cells for the medicine of the future.

REFERENCES

1. K. Bourzac, First Human Test of Optogenetics, *MIT Technology Review*, Feb 19 2016, https://www.technologyreview.com/s/601067/texas-woman-is-the-first-person-to-undergo-optogenetic-therapy/.
2. M. May, Synthetic Biology's Clinical Applications, *Science*, 2015, 15.
3. For some basic information and links to great multimedia materials that explain the nuts and bolts of synthetic biology from one of the first synthetic biology communities see Synberc. https://www.synberc.org/what-is-synbio.
4. G. Baldwin, T. Bayer, P. S. Freemont, T. Ellis, K. Polizzi, G.-B. Stan and R. K. Kitney, *Synthetic Biology: A Primer*, revised edn, Imperial College Press, 2016.
5. D. K. Ro, E. M. Paradise, M. Ouellet, K. J. Fisher, K. L. Newman, J. M. Ndungu, K. A. Ho, R. A. Eachus, T. S. Ham, J. Kirby, M. C. Chang, S. T. Withers, Y. Shiba, R. Sarpong and J. D. Keasling, *Nature*, 2006, **440**, 940–943.
6. C. J. Paddon, P. J. Westfall, D. J. Pitera, K. Benjamin, K. Fisher, D. McPhee, M. D. Leavell, A. Tai, A. Main, D. Eng, D. R. Polichuk, K. H. Teoh, D. W. Reed, T. Treynor, J. Lenihan, H. Jiang, M. Fleck, S. Bajad, G. Dang, D. Dengrove, D. Diola, G. Dorin, K. W. Ellens, S. Fickes,

J. Galazzo, S. P. Gaucher, T. Geistlinger, R. Henry, M. Hepp, T. Horning, T. Iqbal, L. Kizer, B. Lieu, D. Melis, N. Moss, R. Regentin, S. Secrest, H. Tsuruta, R. Vazquez, L. F. Westblade, L. Xu, M. Yu, Y. Zhang, L. Zhao, J. Lievense, P. S. Covello, J. D. Keasling, K. K. Reiling, N. S. Renninger and J. D. Newman, *Nature*, 2013, **496**, 528–532.

7. W. Weber and M. Fussenegger, *Nat. Rev. Genet.*, 2012, **13**, 21–36.

8. Z. Kis, H. A. S. Pereira, T. Homma, R. M. Pedrigi and R. Krams, *J. R. Soc., Interface*, 2015, **12**(106), DOI: 10.1098/rsif.2014.1000.

9. M. Eisenstein, *Nature*, 2016, **531**, 401–403.

10. C. J. Sullivan, E. D. Pendleton, H. H. Sasmor, W. L. Hicks, J. B. Farnum, M. Muto, E. M. Amendt, J. A. Schoborg, R. W. Martin, L. G. Clark and M. J. Anderson, *Biotechnol. J.*, 2016, **11**, 238–248.

11. P. R. Dormitzer, P. Suphaphiphat, D. G. Gibson, D. E. Wentworth, T. B. Stockwell, M. A. Algire, N. Alperovich, M. Barro, D. M. Brown, S. Craig, B. M. Dattilo, E. A. Denisova, I. De Souza, M. Eickmann, V. G. Dugan, A. Ferrari, R. C. Gomila, L. Han, C. Judge, S. Mane, M. Matrosovich, C. Merryman, G. Palladino, G. A. Palmer, T. Spencer, T. Strecker, H. Trusheim, J. Uhlendorff, Y. Wen, A. C. Yee, J. Zaveri, B. Zhou, S. Becker, A. Donabedian, P. W. Mason, J. I. Glass, R. Rappuoli and J. C. Venter, *Sci. Transl. Med.*, 2013, **5**(185), 1–12.

12. S. Young Rojahn, Synthetic biology could speed flu vaccine production, *MIT Technology Review*, May 14 2013, https://www.technologyreview.com/s/514661/synthetic-biology-could-speed-flu-vaccine-production/.

13. M. Amidi, M. de Raad, D. J. A. Crommelin, W. E. Hennink and E. Mastrobattista, *Syst. Synth. Biol.*, 2011, **5**, 21–31.

14. J.-Y. Trosset and P. Carbonell, *Drug Des., Dev. Ther.*, 2015, **9**, 6285–6302.

15. Eurostem cell website has a wealth of information on the potential application of iPSC cells for drug discovery and therapy as background to the applications for synthetic biology. Refer to the factsheet on iPSC cells. http://www.eurostemcell.org/factsheet/reprogramming-how-turn-any-cell-body-pluripotent-stem-cell.

16. Y. M. Yang, S. K. Gupta, K. J. Kim, B. E. Powers, A. Cerqueira, B. J. Wainger, H. D. Ngo, K. A. Rosowski, P. A. Schein, C. A. Ackeifi, A. C. Arvanites, L. S. Davidow, C. J. Woolf and L. L. Rubin, *Cell Stem Cell*, 2013, **12**(6), 713–726.
17. B. Zakeri and T. K. Lu, *ACS Synth. Biol.*, 2013, **2**(7), 358–372.
18. W. Weber, B. N. Link, M. Spielmann, B. Keller, C. C. Weber and M. Fussenegger, Inducible gene expression for antibiotic drug discovery and diagnostics, in *Animal Cell Technology Meets Genomics*, ed. F. Godia and M. Fussenegger, Vol. 2, ESACT Proceedings, Springer, 2005, pp. 345–350.
19. K. Yamanaka, K. Yamanaka, K. A. Reynolds, R. D. Kersten, K. S. Ryan, D. J. Gonzalez, V. Nizet, P. C. Dorrestein and B. S. Moorea, *Proc. Natl. Acad. Sci. U. S. A.*, 2013, **111**(5), 1957–1962.
20. H. Ye, M. Daoud-El, R. W. Peng and M. Fussenegger, *Science*, 2011, **332**(6037), 1565.
21. H. Ye, G. Charpin-El, Hamri, K. Zwicky, M. Christen, M. Folcher and M. Fussenegger, *Proc. Natl. Acad. Sci. U. S. A.*, 2013, **110**(1), 141–146.
22. C. Kemmer, M. Gitzinger, M. Daoud-El Baba, V. Djonov, J. Stelling and M. Fussenegger, *Nat. Biotechnol.*, 2010, **28**(4), 355–360.
23. P. Saxena, B. C. Heng, P. Bai, M. Folcher, H. Zulewski and M. Fussenegger, *Nat. Commun.*, 2016, 7, 11247.
24. L. Warren, P. D. Manos, T. Ahfeldt, Y. H. Loh, H. Li, F. Lau, W. Ebina, P. K. Mandal, Z. D. Smith, A. Meissner, G. Q. Daley, A. S. Brack, J. J. Collins, C. Cowan, T. M. Schlaeger and D. J. Rossi, *Cell Stem Cell*, 2010, 7(5), 618–630.
25. Z. Eshar, *Hum. Gene Ther.*, 2014, **25**(9), 773–778.
26. C. June, *Hum. Gene Ther.*, 2014, **25**(9), 779–784.
27. V. Brower, The CAR T-Cell Race, *The Scientist*, April 1 2015, http://www.the-scientist.com/?articles.view/articleNo/42462/title/The-CAR-T-Cell-Race/.
28. https://www.synlogictx.com/.
29. H. Ando, S. Lemire, D. P. Pires and T. K. Lu, *Cell Syst.*, 2015, **1**(3), 187–196.

Subject Index

References to figures are given in *italic* type. References to tables are given in **bold** type.